Emerson™ Performance Learning Platform

Contents

Section 1: General Information

1.1   Operation and Setup

- See documents XXXXXX and XXXX for more information on operation and setup of the Performance Learning Platform.
- See www.xxxx.com for access to operations manuals.

1.2   Additional Training

- For a video describing the PLP components, go to the following link:

  https://videos.emerson.com/detail/video/5978816530001/

- The intent of this document is to provide the student with enough DeltaV operation and configuration knowledge to perform the labs.   There are detailed DeltaV operation and configuration classes available through Emerson Educational Services should the student require further knowledge.

  For more information on Emerson training classes visit the following website:

  https://www.emerson.com/en-us/automation/services-consulting/educational-services

- For more information on PLP instrumentation go to:

  http://www.emersonstreamingvideo.com/pss/PLP/online/index_home_grid.html?_ga=2.1 96596328.533480057.153886.29433-271996358.1504716980

1.3   General Comments

- The PLP is available in several configurations and multiple flow paths. Labs in sections 2 through 17 were developed on a PLP with a single flow path. Labs in section 18 require a PLP with dual flow paths. If doing the labs in sections 2 through 17 on a dual path PLP unit, simply set module FFIC-300 Out Hi Lim parameter on the Loop Tune Detail to 0%.

- This document was created for use with PLP units that were manufactured in the United States, other codes and standards may apply. Consult your local governing body for specific rules on the installation of equipment and instrumentation in hazardous areas.

Section 2: PLP Infrastructure

2.1 Objectives

When the student has completed this module, the student will:

- Be familiar with the equipment and instruments used on the Performance   Learning Platform (PLP).

- Be familiar with the PLP control system equipment and Inputs and Outputs (I/O).

- Be familiar with Piping and Instrument Diagrams (P&IDs).

- Be able to identify the PLP flow paths.

2.2 Intended Audience

Instrument Technician

Operator

Process Engineer

2.3 Prerequisites

None

2.4 Discussion

For a plant to run smoothly and safely, all personnel in an operating facility need to have a basic understanding of the facility's layout and associated equipment. The governing document that drives all design and operations is the Process and Instrument Diagram (P&ID). It identifies all of the flow paths, process instrumentation and final control elements in the facility. This document must be kept up to date with all changes, per OSHA (Occupational Safety and Health Administration), for the facility to have the right to operate. A basic understanding of P&IDs is critical for operators to communicate operating problems with production management, process engineers and maintenance personnel.

For students to complete future labs, they must become familiar with the PLP infrastructure. There are multiple configurations of the PLP available. Below is a listing of all the PLP instruments, equipment and control system components for a PLP unit with a single flow path.

- Instrumentation

    - FIT-106 – A Rosemount 3051C Differential Pressure transmitter with a 3-valve manifold and 0.63 inch orifice plate

- FIT-206 – A Rosemount 1" 8705 Magnetic flow tube with a Rosemount 8732 integral mount Magnetic flow transmitter with grounding rings

- SV-205 – An Asco 1" 8212A solenoid on/off diaphragm valve

- LIT-101 – A Rosemount 5300 Guided Wave Radar transmitter with 24" probe

- LIT-201 – A Rosemount 5300 Guided Wave Radar transmitter with 24" probe

- LIT-103 – A Rosemount 3051C Differential Pressure transmitter with 3 valve manifold

- FV-106 – A Baumann 1" 24588S threaded 316SS globe valve with 316SS trim and actuator, Cv = 9.5

- FV-206 – A Baumann 1" 24588S threaded 316ss globe valve with 316SS trim and actuator, Cv = 9.5

- LSH-102 – A Rosemount 2120, 24Vdc, Discrete, vibrating fork level switch

- LSH-202 – A Rosemount 2140, 24Vdc Analog, SIL rated, vibrating fork level switch

- PIT-104 – A 3051T Rosemount pressure transmitter with isolation and drain valve manifold

- PIT-204 – A 3051T Rosemount pressure transmitter with isolation and drain valve manifold

- TIT-105 – A 3144P Rosemount temperature transmitter with a 4 wire RTD and thermowell

- TIT-203 – A 3144P Rosemount temperature transmitter with a type J thermocouple and thermowell

- FI-400 – A Brooks ¼" 0 –100 SCFH rotameter with needle valve

- A Regulator ¼" with 0 – 160 PSI pressure gage

- Equipment

  - P-100 – A Dayton centrifugal pump with 1" inlet and ¾" outlet and a 1/3 HP, single phase, 60 Hz, 120VAC, SF 1.75, 56J frame motor

  - P-200 – A Dayton centrifugal pump with 1" inlet and ¾" outlet and a 1/3 HP, single phase, 60 Hz, 120VAC, SF 1.75, 56J frame motor

  - TK-100 – A plastic 30 gallon tank

  - TK-200 – A plastic 30 gallon tank

- Control System Equipment and Software

  - DeltaV PK100 Controller running DeltaV v14.3 software

  - DeltaV operating panel running a Windows 10 operating system

  - 100-DST (Digital Signal Tag) PlantWeb Experience software license with AMS (Asset Management Software) license

- Redundant CIOCs (Charm I/O Card) with mounting base plates

- 11 4-20mA, HART, Analog Input Charms with associated terminal blocks

- 3 4-20mA, HART, Analog Output Charms with associated terminal blocks

- 1 24Vdc, Isolated, Discrete Output Charm with associated relay terminal block

- 1 24Vdc, Isolated, Discrete Input Charm with associated terminal block

- 2 120VAC, High Side, Discrete Output Charms with associated terminal blocks

- 4 120VAC, Isolated, Discrete Input Charms with associated terminal blocks

- Redundant 24Vdc, 10 Amp, power supplies with redundancy module

- Primary and Secondary 10/100BASE-TX switches with RJ-45 connections for Ethernet communications

2.5   Additional Information

For a video describing the PLP components, go to the following link:

https://videos.emerson.com/detail/video/5978816530001/

2.6 Workshop – PLP P&ID Completion

Step 2.6.1

Print out partially completed P&ID in Appendix A.

Step 2.6.2

Fill in the four missing tags and instruments on P&ID.

Step 2.6.3

Using colored highlighters, highlight normal flow paths on P&ID with direction arrows.

2.7   Conclusions

A P&ID is the base document used by production personnel to design, operate, upgrade and maintain a plant. Without an accurate and up to date P&ID, the facility is in violation of its right to operate and it could result in safety issues due to the inability to control flow paths or isolate pipelines properly.

The PLP gives students a chance to familiarize themselves with a real-world control system and industrial grade instrumentation in a safe environment. This exposure will give them a jump start on their careers.

For more information on Emerson training classes visit the following website:

https://www.emerson.com/en-us/automation/services-consulting/educational-services

Section 3: PLP Instrumentation

3.1   Objectives

When the student has completed this module, the student will:

- Understand the key points to consider when specifying instrumentation.

- Be able to identify all the characteristics of an instrument used on the PLP.

- Understand the steps used to maintain and calibrate a piece of instrumentation in a production environment.

3.2   Intended Audience

Instrument Technician

Operator

Process Engineer

3.3   Prerequisites

Section 2 PLP Infrastructure

3.4   Discussion

Instrumentation is the lens from which operations can view what is going on in a production facility. Without instrumentation that is reliable and sized correctly, operations is essentially running blind and this could result in quality and/or safety issues. There are seven key factors to consider when specifying instrumentation.

- Size and type of instrument
- Material of construction (MOC)
- Hazardous area classification requirements
- Installation requirements
- How to verify, calibrate and maintain
- SIS (Safety Instrumented System) rating
- Cost and schedule

3.4.1   Size and Type of Instrument

It is important to know the published accuracy of an instrument. A process engineer can give insight into how critical the accuracy of the measurement needs to be to create quality product. Typically, the closer you get to the lower end of the instrument's published measurement range, the worse the accuracy will be. Sometimes measurements just need to be repeatable but not accurate.

Instruments come in all different sizes and they typically are only good for a narrow range. For example, pressure transmitters can be built with sensor ranges good for -250 to 250 PSI or -25 to 25 INWC. If the wrong one is ordered, you could end up with a paper weight with 4 to 6 weeks needed to correct the situation. Always review the instrument specification before it goes out for bid and review the bid for correctness once it comes back. Vendors are just like the rest of the world; doing more with less. Instrument part numbers can be as long as 25 to 30 characters and it just takes one wrong character to prevent its use or worse yet a safety issue.

It is important to know the characteristics and properties of the fluid, gas or solid you are trying to measure. There are many resources available to obtain information on common chemicals, but in some cases, they need to be calculated by process engineers.

In addition, instrument engineers need to know the temperatures and pressures the instrument will see in normal and abnormal situations to specify it correctly.

3.4.2 Materials of Construction

The instrument's materials of construction must be suited for the process where it will be installed. Instrument corrosion can result in a quality issue or possibly a safety issue. Consider not only the parts of the instrument coming in contact with the process, but also the ambient environment, as sometimes that can lead to reliability issues.

Many vendors don't publish O-ring and seal material of construction in their basic literature. Neoprene is not compatible with a lot of chemical processes and may need to be substituted. Always ask for seal material of construction if not published. Some other material of construction seals might be available if you ask the vendor for a special quotation or you can have special gaskets made if the vendor will give you the size specifications.

3.4.3 Hazardous Area Classification Requirements

In the USA, hazardous area wiring methods as outlined in Article 500 of the National Electric Code are typically done using Divisions, Class and Group classifications although Zones can be used. Different classifications systems are employed throughout the world. Regardless of the hazardous area classification system used, safety personnel, E&I engineers and process engineers should classify and document a process area early on in a project via area classification layouts and elevations. Remember, if the process changes in that area, the hazardous area classification could change and your installed wiring method may not meet your new classification. The instrument must be rated for the area it is being installed in unless being placed in an explosion proof enclosure or purged enclosure. Below is a brief explanation of hazardous area classifications per the United States National Electric Code. Consult your local codes and standards for other areas of the world.

- Class

    - Class I – Flammable gases, flammable liquid produced vapors or combustible liquid-produced vapors

    - Class II – Combustible dust

    - Class III – Ignitable fibers

- Division

    - Division I – Assumes hazard exists under normal operation or frequent maintenance or breakdowns

    - Division II – Assumes hazard exists in abnormal operation or normally exists in insufficient quantity to cause an explosion or combustion

- Group

    - Group A – Acetylene

    - Group B – Gas or vapor that will burn or explode that when mixed with air has a maximum experimental safe gap (MESG) <= 0.45mm or a maximum igniting current ratio (MIC-ratio)<= 0.40. Example: Hydrogen

    - Group C – Gas or vapor that will burn or explode that when mixed with air has a 0.45mm <(MESG) <= 0.75mm or 0.40 <(MIC-ratio) <= 0.80. Example: Ethylene

    - Group D – Gas or vapor that will burn or explode that when mixed with air has a (MESG) > 0.75mm or (MIC-ratio) > 0.80. Example: Propane

    - Group E – Combustible metal dust. Example: Magnesium

    - Group F – Combustible carbon dust. Example: Coal Dust

It is important to know the wiring method that will be used for the installation prior to specifying an instrument. Is the plan for explosion proof equipment wiring or intrinsically safe wiring and I/O?

3.4.4   Installation Requirements

Some instrumentation requires a specific mounting orientation or upstream and downstream straight piping runs to function properly. Also, pay close attention to the ambient conditions. Heat tracing and insulation may be required to prevent freezing. Make sure electronics are isolated from high temperature processes and possibly use remote electronics. If installing in an area with large amounts of electromagnetic interference, shielding may be required.

3.4.5   How to Verify, Calibrate and Maintain

No instrument will last forever. To ensure product quality, an instrument's functionality and calibration must be verified on a periodic basis. The frequency of preventive maintenance and verification should be reviewed and adjusted based on previous verification results and potential for product quality issues. Maintenance could include calibration verification, cleaning or replenishment of consumables such as electrolytes in pH applications. To perform maintenance, calibration and repair safely and while processing, the instrument should be installed with the ability to isolate it, relieve pressure and drain it. Consider flush and drain ports with valves that can be opened slowly and directed away from the individual doing the maintenance. Consider isolating control valves with upstream and downstream isolation valves and bypass lines so the control valve can be pulled during production.

If frequent calibrations are required as with pH sensors, consider where the instrument is mounted and the possibility of seals that allow extraction and isolation under pressure. If instrument sensors are installed in pipe bridges, purchase the instrument with remote mount electronics for ease of access.

On bleed and calibration ports, set them up with fittings that are compatible with the associated NIST (National Institute of Standards and Technology) certified calibration standard.

3.4.6 SIS Rating

Certain instrumentation in a facility might be a part of a Safety Instrumented System. This is determined by conducting a Hazardous and Operability Analysis (HAZOP) and documenting the organization's acceptable risk tolerance. Once the risk tolerance is documented, a LOPA (Layer of Protection Analysis) can be conducted to determine the current Probability of Failure on Demand (PFD). If the PFD is greater than the acceptable risk tolerance, additional SIS instrumentation might be needed. For example, one may need to add additional sensing elements in a two of three voting configuration with SIS rated redundant controllers and multiple final control elements in series. Standard ISA 84 (Instrument Society of America) outlines the procedure for adding, quantifying and maintaining a Safety Instrumented System.

3.4.7 Cost and Schedule

In an ideal world, engineers would not need to worry about cost and schedule. Unfortunately, none of us live in an ideal world. Almost all instruments are made to order. When specifying an instrument, make sure it can be obtained in the time needed to meet the project schedule. One may need to order an instrument prior to everything being completely locked down in order to meet the required instrument lead time. Do this only as a last resort, as it could turn into an expensive mess. Remember the impact of an incorrect instrument is felt long after a budget overruns or deadlines are forgotten.

When evaluating the cost of an instrument, consider the installed cost. If the instrument requires heat tracing, insulation and special tubing for it to function properly, that should be considered in the cost. Also consider plant inventory (spares) in the instrument selection process. For example, if you already stock a similar length level switch made of Hastelloy C276 but 316SS will work, consider using the Hastelloy C276 switch to reduce plant spares.

3.5 Workshop – Instrument Sizing and Materials of Construction Identification

Step 3.5.1

Locate pressure transmitter PIT-104 on the PLP and write down its model number.

Step 3.5.2

Go to a computer with internet access and Google the part number. *You may need to put in the series model to find the correct reference manual.*

Step 3.5.3

Go to the Ordering Information section of the manual and decode the entire part number identifying pressure range size, material of construction and all options.

_____

_____

_____

_____

_____

3.6   Workshop – Instrument Calibration and Maintenance

Step 3.6.1

Document steps to verify PIT-104 calibration as if it were installed in a production system.

_____

_____

_____

_____

_____

_____

_____

_____

_____

Step 3.6.2

Document steps to remove PIT-104 calibration as if it were installed in a production system.

_____

_____

_____

_____

_____

_____

_____

_____

_____

3.7 Conclusions

Instrumentation is the view into the process. There are numerous factors to consider when specifying an instrument and any one of them could cause the instrument to fail or not meet the required quality or precision of the measurement. Specifying the correct instrument for an application requires collaboration between process engineers, instrument engineers, operators and safety personnel.

For more information on Emerson training classes visit the following website:

https://www.emerson.com/en-us/automation/services-consulting/educational-services

Section 4: DeltaV Operate Navigation

4.1 Objectives

When the student has completed this module, the student will:

- Be able to navigate around DeltaV Operate.

- Be able to navigate around Process History View (PHV).

- Be able to use DeltaV Control Studio On-Line.

- Be able to use DeltaV Diagnostics.

- Be able to add point to the DeltaV Continuous Historian.

- Be able to create and save trends in Process History View.

4.2 Intended Audience

Instrument Technician

Operator

Process Engineer

4.3 Prerequisites

Section 2 PLP Infrastructure

4.4 Discussion

All operators, maintenance personnel, process engineers and production supervisors need to be able to navigate around the control system to troubleshoot issues and check production rates. The PLP uses a DeltaV Distributive Control System to control flow rates and levels. The intent of this learning module is to give students a basic understanding of how to maneuver around DeltaV to complete future labs. There are comprehensive DeltaV operator training and DeltaV configuration courses available through Emerson™.

4.4.1   DeltaV Logon

When the system first boots up, it will bring up the FlexLock banner shown below. The purpose of the FlexLock banner is to restrict non-qualified personnel from getting to the Windows Desktop. In addition, it provides the ability to logon to the system and gives users without operating rights the ability to use DeltaV non-operating applications.

Graphic 4-4-1 FlexLock Banner

To logon, click the DeltaV Logon button and enter the Username: EMERSON and Password: DeltaVE1.

Graphic 4-4-2 DeltaV Logon Popup

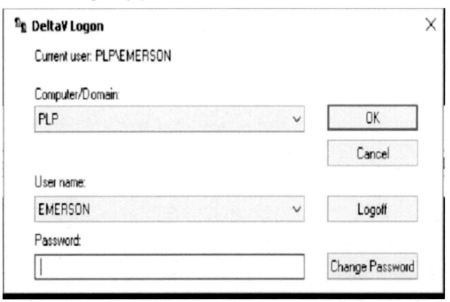

4.5 PLP Main Graphic and Toolbar and Alarm Banner

To bring up the main operating graphic, click on the DeltaV Operate button on the FlexLock banner. By clicking on objects on the screen, you can bring up the faceplates of various DeltaV modules. To get help on the Toolbar and the Alarm Banner, click on the question marks on the upper and lower right corners of the screen.

It is critical to project success and future maintenance that standards are used for configuration and graphic development. The PLP uses Emerson's PCSD (PMO Configuration Standard for DeltaV). The PLP graphics are Gray Scale so that colors such as red and yellow only appear on the graphic if there is something going on which requires the operator's attention. When choosing your graphic color scheme remember some operators are color blind.

Graphic 4-5-1 PLP Main Operating Graphic

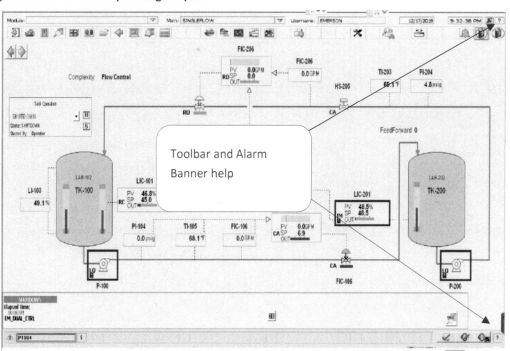

4.5.1   Toolbar Help

Graphic 4-5-2 DeltaV Toolbar and DeltaV Toolbar Help

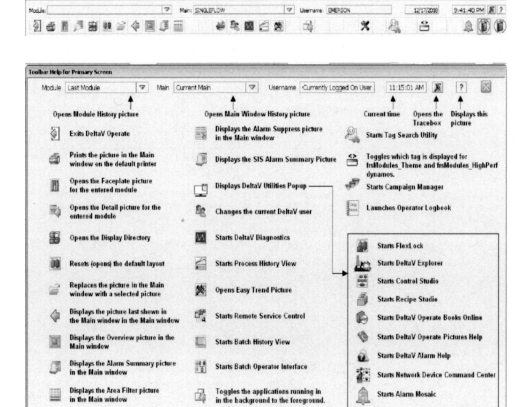

4.5.2   Alarm Banner Help

Unacknowledged alarms will be further to the left with most recent being nearer to the left. Higher priority alarms will be further to the left as grouped by unacknowledged and acknowledged alarms.

Graphic 4-5-3 DeltaV Alarm Banner and Alarm Banner Help

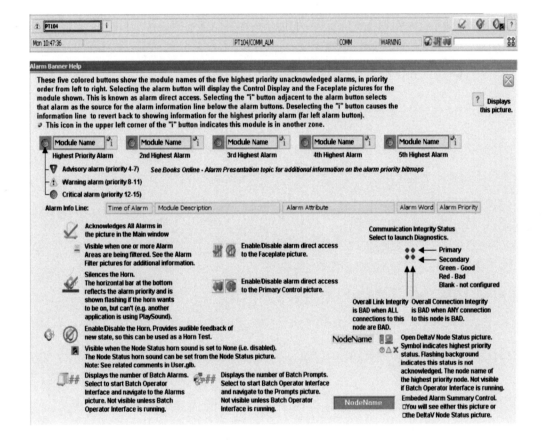

4.6 DeltaV Module Faceplates and Details

There are four main module faceplates available from the main graphic on the PLP.

- Discrete Inputs (DI)
    - Level Switches
- Discrete Control Devices (DCD)
    - Pumps
    - On/Off Valves
- Analog Inputs (AI)
    - Pressure Transmitters
    - Temperature Transmitters
- Loops (PID)
    - Flow Control Loops
    - Level Control Loops

4.6.1   DI Faceplate

Graphic 4-6-1 DeltaV DI Faceplate

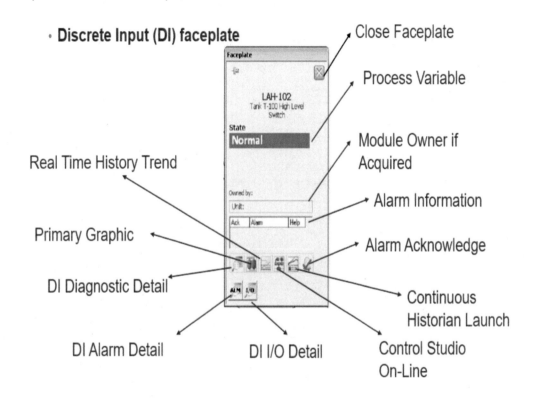

4.6.2   DI Diagnostic Detail

This detail is used to troubleshoot I/O and programming problems. It serves as a good indication that a module is functioning as expected. Be careful though — unless the discrete input channel is configured with open loop monitoring, a cut wire will behave just like a low (zero) input.

Graphic 4-6-2 DeltaV DI Diagnostic Detail

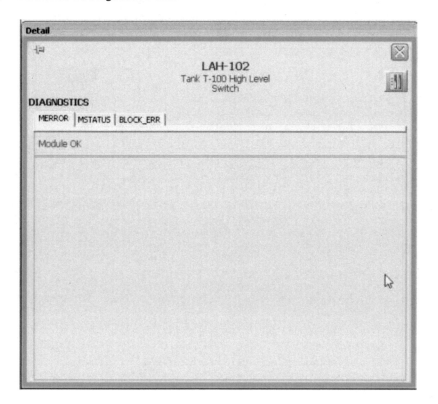

4.6.3   DI I/O Detail

This detail is used to simulate inputs for code testing or bypassing instrument problems. It also allows the instrument technician to monitor the field value. Sometimes instruments fail and it may be worse to sit in a process step and wait for repair versus moving along in the sequence. Simulation can be used to provide the required input so the program advances. Use this function with great care and always report the instrument failure to maintenance for repair. Simulation of an instrument could create an unsafe situation, so always log it so others know the input is not tracking the field value.

Graphic 4-6-3 DeltaV DI I/O Detail

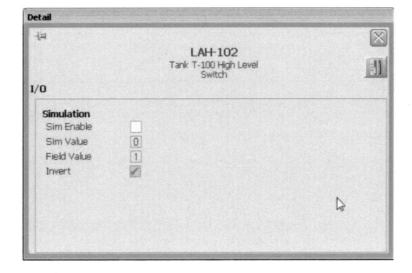

4.6.4   DI Alarm Detail

This detail is used to enable and disable alarms.

- A Discrete Alarm is activated if the State of the Discrete Input matches the value of the Discrete Limit

- A Module Alarm is active when there is an input failure or when the process value status is bad

Only enable alarms which require operator action. Too many nuisance alarms will result in operators ignoring them. If bypassing an alarm, log it and review that log at the beginning of each shift. Alarms give the operator the ability to avoid process upsets, so by disabling them you can increase your chances of creating out of spec product or unsafe operating conditions.

Graphic 4-6-4 DeltaV DI Alarm Detail

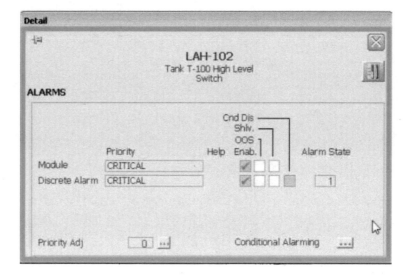

4.6.5   DCD Faceplate

DCDs can only have one passive state but can have multiple active states. An example would be a two-speed motor. The passive state would be STOPPED with active states SLOW and FAST. On the DCD faceplate, the top requested setpoint state is the passive state. In this case STOP.

Graphic 4-6-5 DeltaV DCD Faceplate

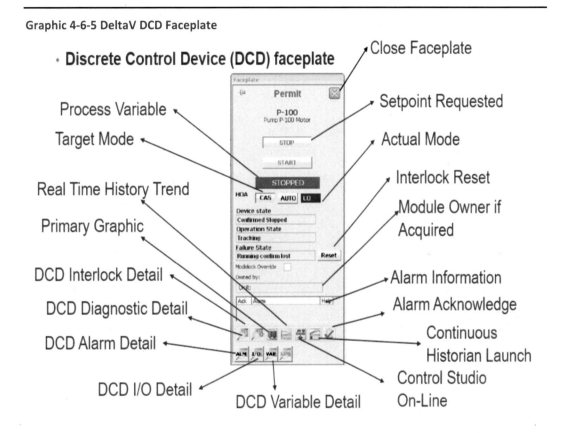

4.6.6   DCD Diagnostic Detail

The DCD Diagnostic Detail is used to troubleshoot I/O and programming problems.

Graphic 4-6-6 DeltaV DCD Diagnostic Detail

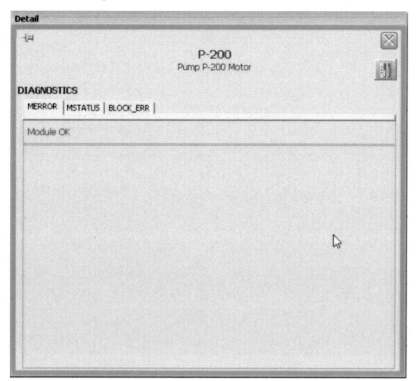

4.6.7 DCD Alarm Detail

The DCD Alarm Detail is used to enable and disable alarms. The alarms for a DCD are defined as follows.

- DCD Fail Alarm

    - Fail Alarm is activated when an **(interlock occurs)** or a **(write alarm occurs)**. A Fail Alarm can be reset through the faceplate even if the interlock is not cleared.

    - Typical DCD fail alarms come from MCC buckets being locked out, limit switch problems on valves or the instrument air supplying a valve actuator being turned off.

- DCD Module Alarm

    - A Module Alarm is activated when there is an input failure **(device failed)**, or when the process value status is bad **(out of range).**

- DCD Write Alarm (WA)

    - In Auto mode or Local Override, a Write Alarm is triggered when change in requested setpoint is different from current setpoint. Module Alarm and Sentinel Status are combined to FAILURE flag to propagate to higher level.

    - A typical Write Alarm occurs when a valve is interlocked and a sequencer tries to open it.

Graphic 4-6-7 DeltaV DCD Alarm Detail

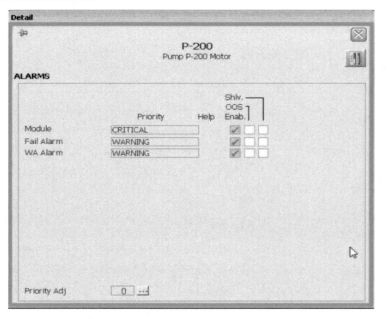

4.6.8 DCD I/O Detail

The DCD I/O Detail is used to simulate DCD State for code testing, or bypassing limit switch and wiring problems. It also gives the operator the ability to extend state confirm times if a valve is sticking. If the confirm time is shorter than the actual time it takes a device to change states, this will result in a fail alarm. If this starts happening on a regular basis, it is usually an indication that maintenance needs to be performed on the device since it is sticking. DeltaV allows you to have individual confirm times for each state.

Graphic 4-6-8 DeltaV DCD I/O Detail

4.6.9   DCD Variable Detail

The DCD Variable Detail gives personnel with the correct privileges the ability to change variables used in the configuration of the module. If code is setup with variables, this minimizes the need to have process control engineers available for minor limit adjustments.

Graphic 4-6-9 DeltaV DCD Variable Detail

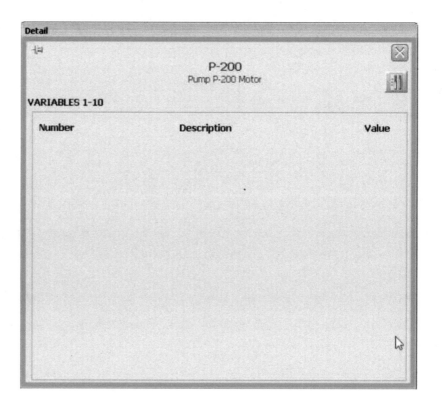

4.6.10   DCD Interlock Detail

The DCD Interlock Detail is used to determine if an interlock, force setpoint or permissive interlock is active on the device and to bypass the condition if necessary. Without this information operators would have no way of determining why a unit shut down, why a valve will not open or why a pump will not start. If you bypass any interlock, force setpoint or permissive interlock, be sure to log it. It may require a change control to bypass an interlock since it could result in an unsafe condition or equipment failure.

Graphic 4-6-10 DeltaV DCD Interlock Detail

4.6.11　AI Faceplate

Graphic 4-6-11 DeltaV AI Faceplate

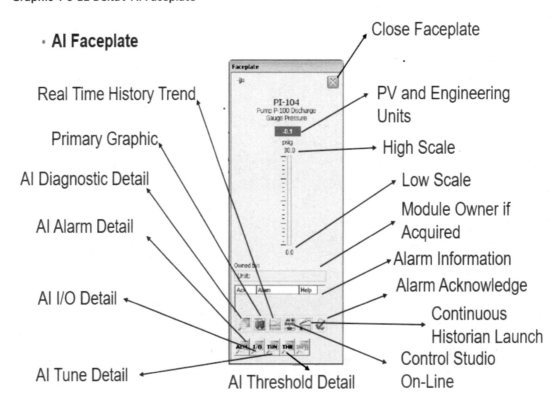

4.6.12   AI Diagnostic Detail

The AI Diagnostic Detail is used to troubleshoot I/O and programming problems. It is a good indication that the module I/O and code are functioning as expected.

Graphic 4-6-12 DeltaV AI Diagnostic Detail

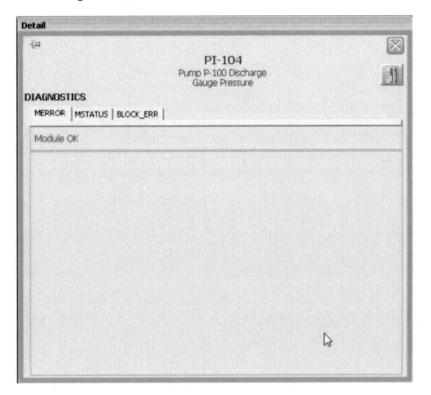

4.6.13   AI Alarm Detail

The AI Alarm Detail is used to enable and disable alarms and change alarm thresholds if needed. Always log alarms that are disabled as this could result in an unsafe condition. DeltaV has the ability to do conditional alarming to prevent nuisance alarms. For example, if the pump feeding a flowmeter is off, it does not make sense to have the low flow alarm active.

- AI High High Alarm

    - A High High Alarm is activated when the Process Value is higher than the High-High Limit

- AI High Alarm

    - A High Alarm is activated when the Process Value is higher than the High Limit

- AI Low Alarm

    - A Low Alarm is activated when the Process Value is lower than the Low Limit

- AI Low Low Alarm

    - A Low Low Alarm is activated when the Process Value is lower than the Low-Low Limit

- AI Module Alarm

    - A Module Alarm is activated when there is an input failure **(device failed)**, or when the Process Value status is bad **(out of range, range)**

- AI Rate Alarm

    - A Rate Alarm is activated when the absolute value of the Process Variable Rate of Change is greater than the Rate of Change limit

Graphic 4-6-13 DeltaV AI Alarm Detail

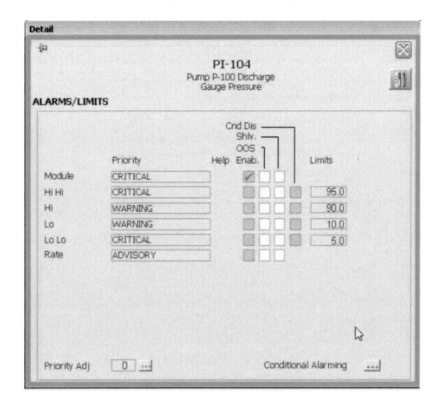

4.6.14   AI I/O Detail

The AI I/O Detail is used to simulate inputs for troubleshooting code problems or bypassing instrument issues. Simulation can be used to push a sequence along if there is an instrument problem. Always log when simulation is used and alert maintenance that there is an instrument problem. Simulation should be used with great care as it could result in an unsafe condition.

Graphic 4-6-14 DeltaV I/O Detail

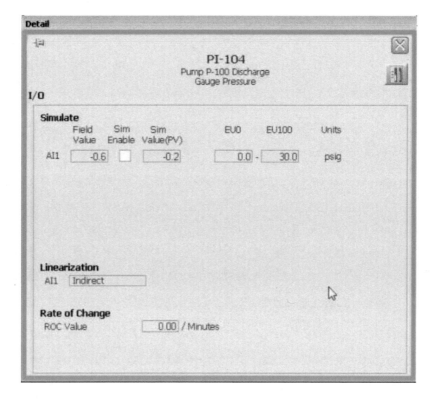

4.6.15   AI Tune Detail

The AI Tune Detail is used to add filtering to a noisy analog input or set the low cutoff value so a good zero is maintained when the device is not in service. Too much filtering can lead to bad control since you can miss true process disturbances. Making the low cutoff too high can mask true operating conditions as well.

Graphic 4-6-15 DeltaV AI Tune Detail

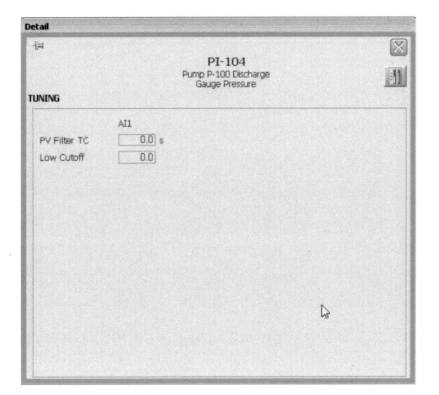

4.6.16   AI Threshold Detail

The AI Threshold Detail is used to enable and disable thresholds. When thresholds are enabled, this give the controls engineer the ability to add smart alarms.

Graphic 4-6-16 DeltaV AI Threshold Detail

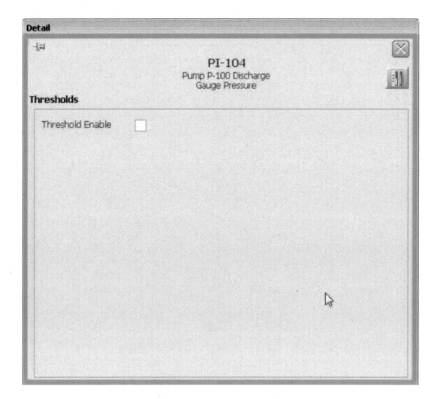

4.6.17 Loop Faceplate

Graphic 4-6-17 DeltaV Loop Faceplate

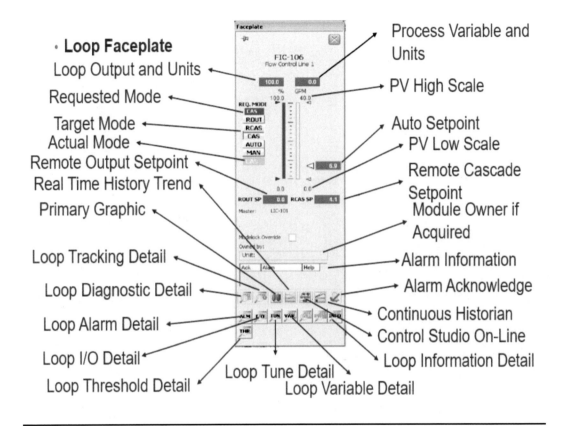

4.6.18   Loop Diagnostic Detail

The Loop Diagnostic Detail is used to troubleshoot I/O and programming problems. It is a good indication that a module is functioning as expected.

Graphic 4-6-18 DeltaV Loop Diagnostic Detail

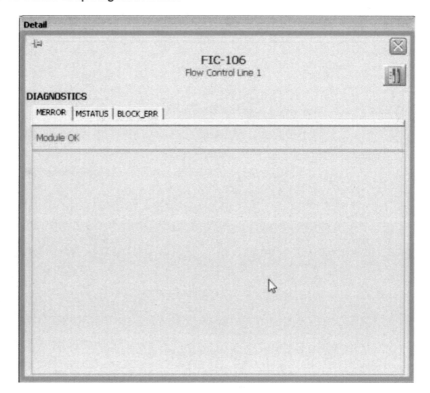

4.6.19   Loop Alarm Detail

The Loop Alarm Detail is used to enable and disable alarms and change alarm thresholds if needed. This detail allows operators to disable alarms or change the alarm thresholds. Always log alarms that are disabled as this could result in an unsafe condition. Remember, too many alarms are equally as bad as no alarms because operators will start to ignore them.

- PID High High Alarm

    - A High High Alarm is activated when the Process Value is higher than the High-High Limit

- PID High Alarm

    - A High Alarm is activated when the Process Value is higher than the High Limit

- PID Low Alarm

    - A Low Alarm is activated when the Process Value is lower than the Low Limit

- PID Low Low Alarm

    - A Low Low Alarm is activated when the Process Value is lower than the Low-Low Limit

- PID Module Alarm

    - A Module Alarm is activated when there is an input failure **(device failed)**, or when the Process Value status is bad **(out of range, range)**

- PID Hi Dev Alarm

    - A Hi Dev Alarm is activated when the Process Variable is greater than the Working Setpoint by the High Deviation Limit

- PID Lo Dev Alarm

    - A Lo Dev Alarm is activated when the Process Variable is less than the Working Setpoint by the Low Deviation Limit

Graphic 4-6-19 DeltaV Loop Alarm Detail

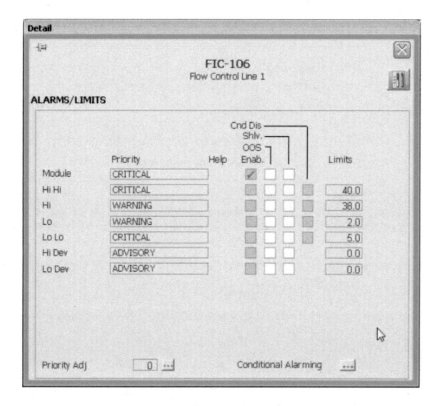

4.6.20 Loop I/O Detail

The Loop I/O Detail is used to simulate inputs for troubleshooting code or bypassing instrument issues. Simulation can be used to push a sequence along if there is an instrument problem. Always log when simulation is used and alert maintenance that there is an instrument problem. Simulation should be used with great care as it could result in an unsafe condition.

Graphic 4-6-20 DeltaV Loop I/O Detail

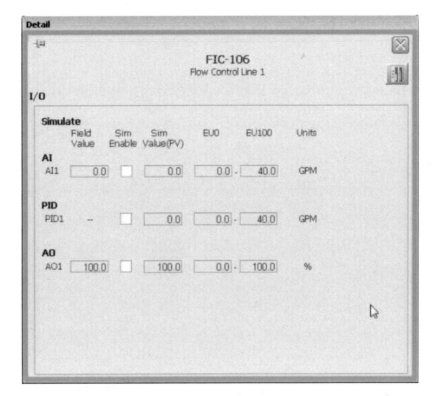

4.6.21   Loop Tune Detail

The Loop Tune Detail is used to modify the loop tuning parameters, loop output limits or loop setpoint limits.

Graphic 4-6-21 DeltaV Loop Tune Detail

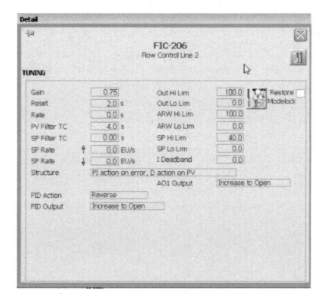

4.6.22   Loop Threshold Detail

The Loop Threshold Detail is used to enable and disable the loop threshold. If thresholds are enabled, this turns on smart alarming.

Graphic 4-6-22 DeltaV Loop Threshold Detail

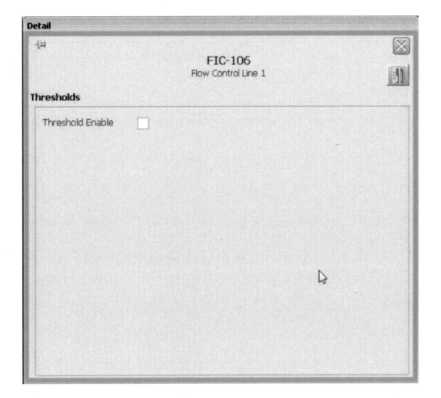

4.6.23   Loop Variable Detail

The Loop Variable Detail gives personnel with the correct privileges the ability to change variables used in the configuration of the module. If code is setup with variables, this minimizes the need to have process control engineers available for minor limit adjustments.

Graphic 4-6-23 DeltaV Loop Variable Detail

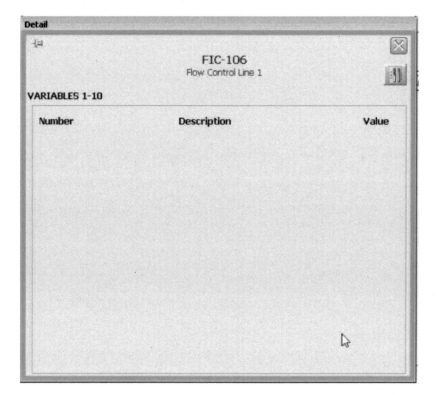

4.6.24   Loop Information Detail

When using a differential pressure transmitter for flow, it requires pressure and temperature compensation to account for temperature and pressure swings in the process fluid. This detail gives the process engineer the ability to modify the standard pressure and temperature conditions if necessary without the need of a controls engineer.

Graphic 4-6-24 DeltaV Loop Information Detail

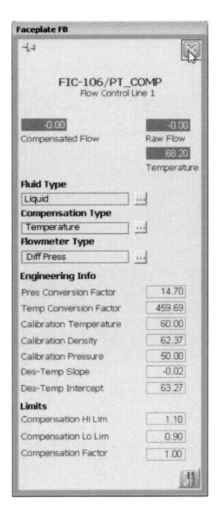

4.6.25   Additional Information

DeltaV Books On-Line is available from the DeltaV Operate Toolbar and does an excellent job of explaining DeltaV applications. For interactive training, consider taking a class offered by Emerson™ Educational Services.

https://www.emerson.com/en-us/automation/services-consulting/educational-services

4.7 DeltaV Process History View (PHV)

4.7.1 DeltaV Process History View Navigation

All operators, engineers and instrument mechanics need to be able to create and modify trends for troubleshooting and monitoring operations. DeltaV has the ability to create custom trends for future use. The Process History View application can trend any DeltaV parameter in real time, but if you want to record the data, you need to configure the point in DeltaV Explorer.

Graphic 4-7-1 DeltaV Process History View Navigation

4.7.2 Custom Trend Creation

Each custom trend can have up to 8 parameters on it. Not all users will have the ability to save process trends for security reasons. Once the trend is saved, it will only be available on that console until a controls engineer is notified and they distribute the new trend to all of the operating consoles.

4.7.2.1 Create a New Custom Trend Using Other Trends

- Open Process History View (PHV)from the DeltaV Toolbar

- Click File, New

Graphic 4-7-2-1 Saving a New Trend

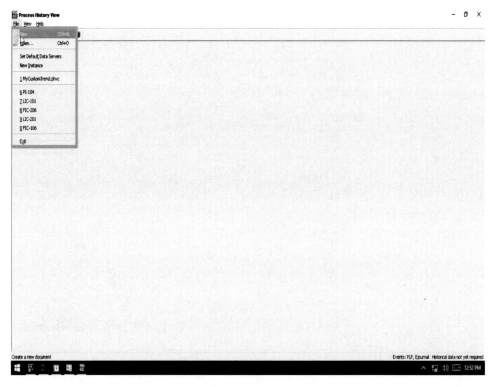

- Choose the type of trend you want to create

  - Select E+Chart for trends that display events and process values (Event examples would be alarms, Setpoint Changes, Operator Opening a Valve, etc.)

  - Select Chart for trends that display process values only

  - Event only displays events and should not be used very often

Graphic 4-7-2-2 Types of Trends

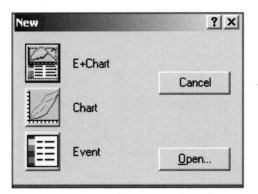

- Enter the Title and Subtitle of the custom trend
  - Click OK

Graphic 4-7-2-3 How to Name a Trend

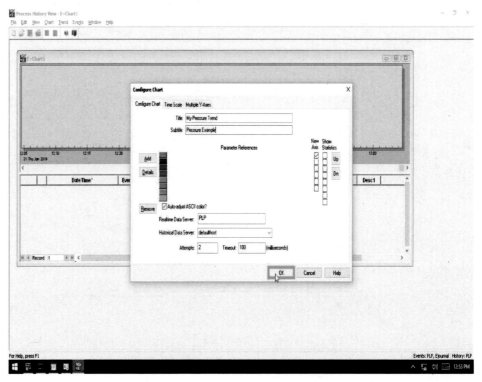

- Save Trend

  - Select File, Save As

  - Save new custom trend with unique name

- Add parameters to the new trend

  - There are multiple ways to add parameters to the new custom trend, but the simplest way will be demonstrated.

- Minimize DeltaV PHV

- The following steps are used to demonstrate how to copy parameters from existing faceplate trends rather than defining them manually. This allows a novice DeltaV user to build trends with little DeltaV expertise.

  - Open DeltaV Operate if it is not already open

  - Open the Faceplate for the first point you want on your custom trend

  - Open Process History View

  - This opens PHV with the faceplate trend for that point.

–   Select Chart, Configure Chart from the PHV Toolbar

Graphic 4-7-2-4 Chart Configuration Popup

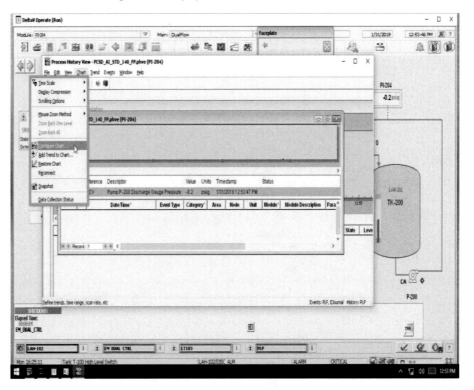

- Select the parameter you want to copy and open up the Configure Trend configuration box

- Copy the contents of the parameter Reference Box

- Close Configure Trend menu, Close Configure Chart Menu, and Close the faceplate trend

Graphic 4-7-2-5 Copy Path of Existing Trend

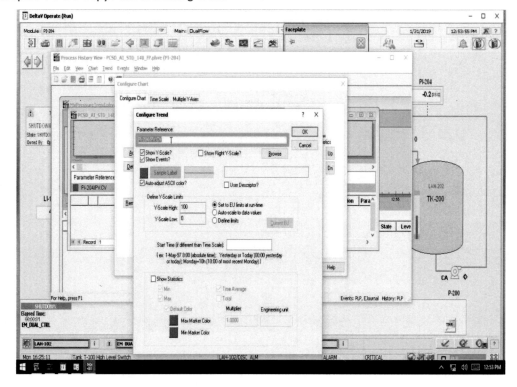

- Open your Custom Trend if not already open in PHV

- Open the Configure Chart menu

  - Select Chart from the PHV Toolbar, Configure Chart

  - Copy the unique parameter reference you copied from the prebuilt trend to your custom trend

  - Change the color of trend line as desired

  - Unless a custom engineering unit scale is desired, leave all other options as the default

  - Select OK to accept changes on the Configure Trend and Configure Chart Menus

Graphic 4-7-2-6 Modify Custom Trend

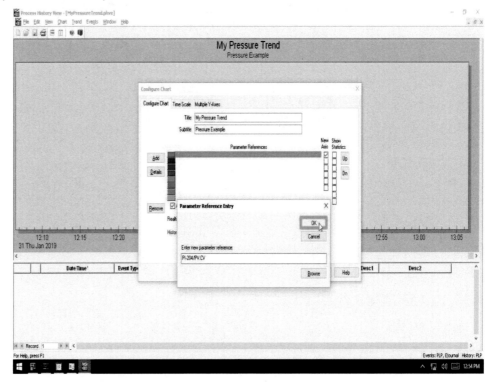

- Each Custom Chart can have a unique Time Scale. The Time Scale is the default amount of time the chart will display when opened.

    – Example: If you typically want to look at the trend over a 6 hour period, set the default Time Scale to 6.

Graphic 4-7-2-7 Adjusting Time Scale

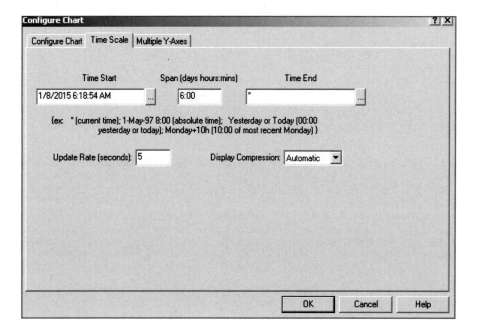

- To open a custom trend once it has been created.

    – Open PHV and select File, Open and then browse to the Custom Trend by name under D:\DeltaV\DvData\Charts

Graphic 4-7-2-8 Opening an Existing Custom Trend

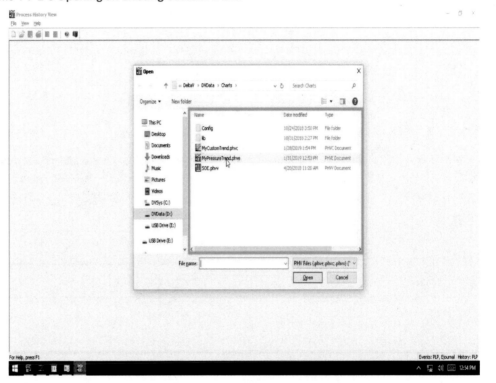

4.8 Workshop – Using Control Studio On-Line

Step 4.8.1

Open DeltaV Operate via the FlexLock Banner.

Step 4.8.2

On the PLP graphic, click on PI-104 to bring up the module faceplate.

Step 4.8.3

Go to the I/O cabinet and remove Charm CHM1-05.

Step 4.8.4

Document what happens to the graphic, faceplate and diagnostic detail for PI-104. Open PI-104 with On-Line Control Studio using the Icon at the bottom of the PI-104 module. What happened to the OUT parameter on the AI block?

_____

_____

_____

Step 4.8.5

Open PI-104 with On–Line Control Studio using the Icon at the bottom of the PI-104 module and click on the AI block. In the lower left window, browse to the OUT parameter. Record the status of the OUT parameter of the AI block by right clicking on the OUT parameter and selecting Set Value.

Step 4.8.6

Re-insert CHM1-05.

4.9 Workshop – Using the DeltaV Continuous Historian

Step 4.9.1

Open DeltaV Operate via the FlexLock Banner.

Step 4.9.2

Open DeltaV Explorer from the ToolBar Icon and in the left window expand the DUALFLOW area under System Configuration/Control Strategies.

Graphic 4-9-2 DeltaV Explorer Area

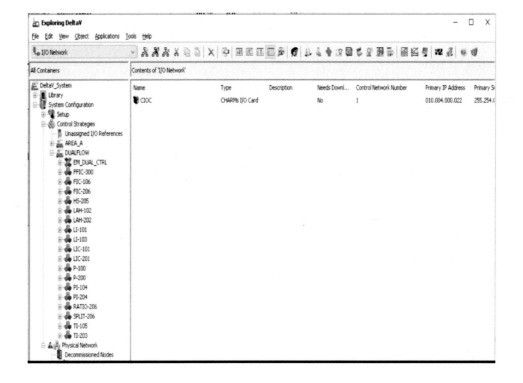

Step 4.9.3

Right click on module PI-104 in the left window and launch History Collection.

Graphic 4-9-1 DeltaV Point Addition

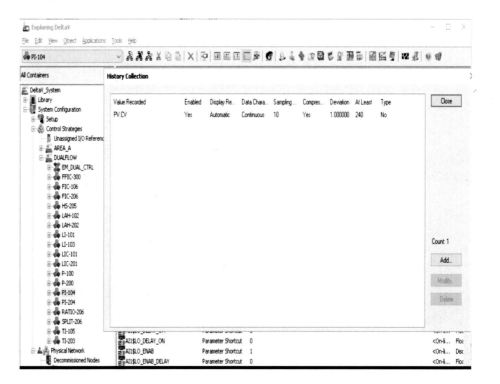

Step 4.9.4

Click the Add button and browse (by double clicking) to path AI1/PV.ST and click OK to add the PV (process variable) status to the Continuous Historian.

Graphic 4-9-4 DeltaV Historian Point Path

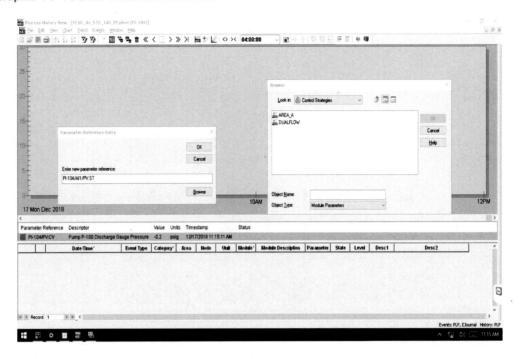

Step 4.9.5

In DeltaV Explorer in the left window, right click on PI-104 and download the module.

Graphic 4-9-5 DeltaV Module Download

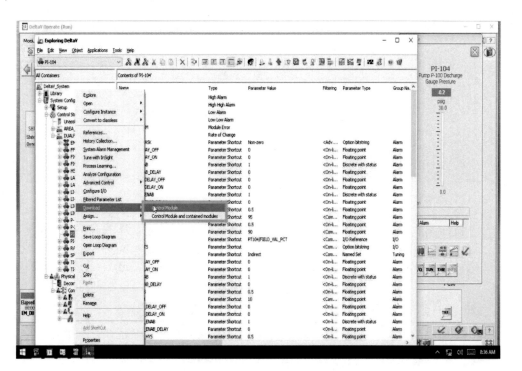

Step 4.9.6

In DeltaV Explorer in the left window, expand the Physical Network and expand the Control Network. Right click on the PLP device and select download ProfessionalPlus Station. The Continuous Historian will now begin recording the PV status.

Graphic 4-9-6 DeltaV Professional Plus Download

Step 4.9.7

Go to DeltaV Operate via the FlexLock Banner.

Step 4.9.8

Click on PI-104 to bring up the module faceplate.

Step 4.9.9

Launch the Continuous Historian from the Icon on the bottom of the PI-104 faceplate.

Step 4.9.10

Add the PI-104 AI1/PV.ST parameter to the trend by clicking the AddTrend Icon on the Continuous Historian Toolbar and browsing (by double clicking) to path DUALFLOW/PI-104/AI1/PV/ST and selecting OK.

Step 4.9.11

Go to the I/O cabinet and remove the wire on terminal 1 of Charm CHM1-05.

Step 4.9.12

Record what happens to the PV and PV status.

_____

_____

_____

Step 4.9.13

Re-install the wire on terminal 1 of Charm CHM1-05.

4.10 Workshop – Using DeltaV Diagnostics

Step 4.10.1

Open DeltaV Operate via the FlexLock Banner.

Step 4.10.2

Open DeltaV Diagnostics via the DeltaV Operate Toolbar.

Step 4.10.3

In the left window expand I/O Network/CIOC/Charms and click on CHM1-05.

Step 4.10.4

Record the status and value of CHM1-05 in the upper right window.

_____

_____

_____

Step 4.10.5

Go to the I/O cabinet and remove Charm CHM1-05.

Step 4.10.6

Record the status of CHM1-05 in the upper right window.

_____

_____

_____

Step 4.10.7

Re-insert Charm CHM1-05.

4.11   Conclusions

There are numerous applications available within DeltaV which can assist operators, maintenance personnel and process engineers in diagnosing problems. The preceding exercises were designed to expose the student to some of these tools. If an in depth understanding of DeltaV applications is desired, Emerson™ offers a wide variety of classes.

For more information on Emerson training classes visit the following website:

https://www.emerson.com/en-us/automation/services-consulting/educational-services

Section 5: DeltaV Modes and Signal Status

5.1   Objectives

When the student has completed this module, the student will:

- Understand DeltaV modes.

- Understand the purpose of DeltaV signal status.

- Understand DeltaV PID control options.

- Understand the effect on mode on control.

- Understand how signal status is propagated.

5.2   Intended Audience

Instrument Technician

Operators

Process Engineer

5.3   Prerequisites

Section 2 PLP Infrastructure

Section 4 DeltaV Operate Navigation

5.4   Discussion

5.4.1   DeltaV Modes

A basic understanding of operating modes is critical for the safe and efficient operation of a production facility. The primary function of an operator is to take manual control of a facility if the control system is producing out of spec product. The effect of a mode change must be understood before it is executed, or it could result in creating unintended consequences. Usually in a continuous operation, if an operator needs to put a loop in manual mode, it is the result of an unanticipated instrument problem and it must happen quickly to prevent a shutdown or quality issues. Each control system vendor has their own modes, however; there are seven modes in DeltaV.

- MAN (Manual)

- AUTO (Automatic)

- ROUT (Remote Output)

- RCAS (Remote Cascade)

- CAS (Cascade)

- LO (Local Override)

- IMAN (Initialization Manual)

5.4.1.1   MAN

- Operator is in control of the function block's output. In the case of the PID function block, the operator is setting the loop output. A simple example is your car's cruise control and the driver has control of the throttle.

- Applies to PID and analog function blocks.

- If in MAN mode, the setpoint value can track the process variable if selected in the PID block CONTROL_OPTS. This allows for smooth transfer between modes.

- A PID block has a Target mode and Actual mode. If the PID block Actual mode is in LO (Local Override), the output will go to the TRK_IN value.

5.4.1.2   AUTO

- Applies to Discrete Control Devices (DCD) and PID function blocks.

- Operator is in control of the function block's setpoint. In the case of the PID function block, the operator is setting the loop setpoint and the PID function block will adjust the loop output to maintain the desired setpoint based on the loop tuning settings (Gain, Reset and Rate). A simple example is your car's cruise control and it is controlling at a driver supplied setpoint.

- In the case of the DCD function block, the operator is requesting the device state.

- Just because a setpoint request is made, it does NOT mean the loop will achieve setpoint or the device will achieve the requested setpoint or state. The tuning could be wrong, or the loop could be interlocked and be tracking. In the case of a DCD, the air to the valve could be off.

- Setpoint velocity limits apply in AUTO mode.

5.4.1.3   ROUT

- The control system is in control of the function block's output. The control system is setting the loop output on the PID function block to the ROUT_IN value. A simple example is a self-driving car's cruise control where the computer is setting the throttle to a specific percent open.

- Applies to the PID function block.

- The ROUT_IN value could come from a Fixed Sequence, a Batch Phase or some other function block.

- A typical use of ROUT mode is to head start a valve to a position close to its final control position prior to putting the loop in RCAS mode.

- If in ROUT mode, the setpoint value can track the process variable if selected in the PID block CONTROL_OPTS. This allows for smooth transfer between modes.

5.4.1.4   RCAS

- The control system is in control of the function block's setpoint. The control system is setting the loop setpoint on the PID function block and the PID function block will adjust the loop output to maintain the desired setpoint based on the loop tuning settings (Gain, Reset and Rate). A simple example is a self-driving car's cruise control where the computer is setting the cruise control speed setpoint based on a speed limit sign.

- Applies to the PID function block.

- The RCAS_IN value could come from a Fixed Sequence, a Batch Phase or some other function block.

- Just because a setpoint request is made, it does NOT mean the loop will achieve setpoint. The tuning could be wrong, or the loop could be interlocked and be tracking.

5.4.1.5   CAS

- Another PID block is in control of the function block's setpoint. A Master PID block is setting the Slave PID block's setpoint. The Slave PID function block will adjust the loop output to maintain the desired setpoint coming from the Master based on the loop tuning settings (Gain, Reset and Rate). An example would consist of two control loops on a self-driving car. A master distance loop trying to keep a constant distance between the driver's car and the car in front of the driver and a Slave speed loop controlling speed. The distance loop output would adjust the speed loop setpoint to maintain a constant distance. This can only be done within the constraints of the maximum and minimum speed setpoints under normal conditions.

- In the case of a DCD, the remote setpoint comes from a Fixed Sequence or Batch Phase.

- Applies to the PID and DCD function blocks.

- The CAS_IN value comes from another loop.

- Just because a CAS setpoint value is being sent from the Master loop, it does NOT mean the Slave loop will achieve setpoint or even accept the setpoint from the Master loop. The tuning could be wrong, or the loop could be interlocked and be tracking.

- The Master loop output will track the Slave loop's setpoint when the Slave loop is not in CAS mode to allow for smooth transfer to CAS mode. The Master loop will go to IMAN mode, if the Slave loop is not in CAS mode.

5.4.1.6   LO

- Applies to the DCD and PID function blocks and cannot be selected by the operator or sequence code.

- The PID block is tracking and the loop output is set to the TRK_IN value.

- In the case of a DCD block, the output state is locked in the passive state regardless of the requested state (setpoint).

- LO mode is used primarily for interlocking or overriding normal analog control. No matter what mode, output or setpoint the operator or computer selects, the discrete output is locked in the passive state or the loop is tracking the TRK_IN value until the LO condition clears. On PIDs, there is an exception for MAN mode if configured in CONTROL_OPTS. LO mode is used to prevent operators or sequence code from getting the process into an unsafe situation. They are always active.

- Typically for a loop, assuming the Target mode remains in CAS, RCAS or AUTO, the loop will resume control after the tracking condition clears. However, this is configurable.

- Typically for a DCD, the device will remain in the passive state until the device is reset via the operator or sequence code. However, this is configurable with DEVICE_OPTS.

5.4.1.7 IMAN

- The PID block is unable to control in MAN, AUTO, RCAS, ROUT or CAS because there is a problem downstream or a Slave PID function block is not in CAS mode.

- Applies to the PID function block and cannot be selected by the operator or sequence code.

- IMAN is typically seen on a Master loop when a Slave loop in a Cascade configuration is not in CAS mode. This will prevent the Master loop from winding up its output when it really doesn't have control.

5.4.2 DeltaV Signal Status

Older control systems and many Programmable Logic Controllers (PLCs) do not have status of the signal. Only the value of a signal is transmitted. This means if you want to take specific actions on a transmitter's bad signal or prevent a loop from winding up if a transmitter saturates, this must be programmed by the integrator and at an additional cost. Typical uses of signal status are to reduce or eliminate the integral action on a Master loop if the Slave loop is at setpoint limits or to freeze an output of a loop if a transmitter fails. All of this is extremely important in a continuous operation to prevent a shutdown or a safety issue. From a quality standpoint, you may choose to shut down a batch process if you have bad transmitter quality as denoted by the signal status. This could prevent an overcharge situation resulting in out of spec product.

The majority of DeltaV function blocks will automatically propagate signal status to the upstream function block, but be careful with CALC function blocks, as they do not propagate signal status. Loop status is propagated upstream via the BKCAL_OUT parameter. This is critical for good cascade, ratio, override or split range control. There are 227 DeltaV status codes. All the status codes are listed in DeltaV Books Online in the Index under Function Block Status Values.

Graphic 5-4-2 DeltaV Function Block Status Values

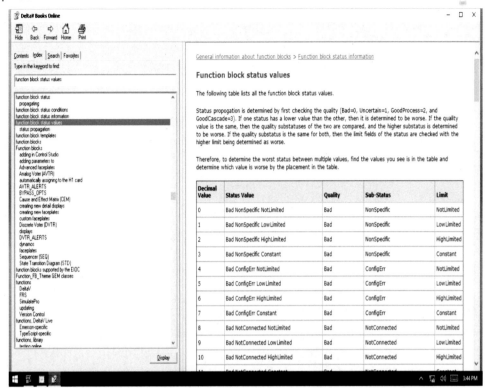

5.6 Workshop – Effect of Mode on Control

Step 5.5.1

Open DeltaV Operate via the FlexLock Banner.

Step 5.5.2

From the graphic screen, set the Complexity to Flow Control and Tank Operation to STARTUP. PLP will startup. Allow FIC-106 and FIC-206 to stabilize at around 5.0 GPM. *You may need to run Fill Tank 1 first to get TK-100 above 55%.*

Graphic 5-5-2 PLP Startup

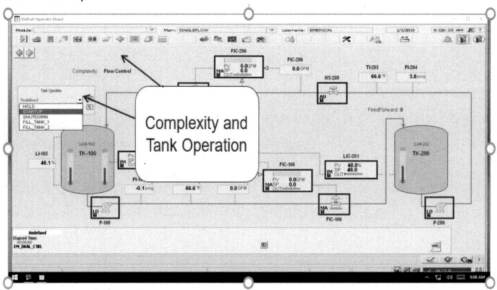

Step 5.5.3

From the graphic screen, click on FIC-206 to bring up the loop faceplate.   Place FIC-206 in MAN mode and launch Process History View from the Icon on the bottom of the faceplate.

Step 5.5.4

Pinch off manual valve downstream of FV-206 and record what happens to the setpoint and PV of the FIC-206.

_____

_____

_____

Step 5.5.5

With the manual valve pinched back to about 4.5 GPM on FIC-206, place FIC-206 in AUTO mode with a setpoint of 5.0 GPM. Record what happens to the loop output. At the time the loop was put in AUTO, what was the setpoint? Why is this important?

_____

_____

_____

Step 5.5.6

Place FIC-206 in RCAS mode. Does the setpoint stay at 5.0 GPM? Record what happens if you pinch back on the manual valve downstream of FIC-206.

_____

_____

Step 5.5.7

For 5 seconds, place FIC-206 in ROUT mode and then return to RCAS mode. Record what happens to the loop FIC-206 output and setpoint.

_____

_____

Step 5.5.8

From the operating graphic, open faceplate for loop FIC-106 and make sure it is in CAS mode. Wait until the flow on FIC-106 and FIC-206 are at about 5.0 GPM. Note the output of LIC-101 is the setpoint of FIC-106, and it is adjusting the flow rate on FIC-106 to maintain a constant level in TK-100.

Step 5.5.9

Place FIC-206 in AUTO mode with a setpoint of 4.5 GPM and record what happens to the flow rate on FIC-106.

_____

_____

Step 5.5.10

Place FIC-206 in RCAS mode and allow the system to stabilize for 5 minutes with FIC-106 and FIC-206 at about 5.0 GPM.

Step 5.5.11

Place FIC-106 in AUTO with a setpoint of 4.5 GPM. Record what happens to Actual mode and output of LIC-101.

_____

_____

Step 5.5.12

Place FIC-106 in RCAS mode and allow the system to stabilize for 5 minutes with FIC-106 and FIC-206 at about 5.0 GPM.

Step 5.5.13

From the operating graphic, bring up the faceplate of LAH-202 and open the I/O detail.

Step 5.5.14

Simulate a High level by clicking Sim Enable and setting the Sim Value to 0.   Record what happens to the Actual mode and output of FIC-106.

Graphic 5-5-14 High Level Switch Simulation

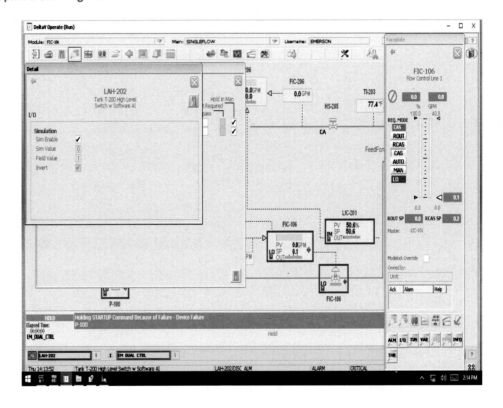

Step 5.5.15

Remove the High level simulation on LAH-202 by unclicking the Sim Enable box and restart the unit by setting the Tank Operation to STARTUP. Allow flows to stabilize around 5.0 GPM.

5.7 Workshop – Propagation of DeltaV Signal Status

Step 5.6.1

From the graphic screen, click on FIC-106 to bring up the loop faceplate.   Launch Control Studio On-Line and record the status of the AI1 block OUT.

Graphic 5-6-1 FIC-106 Status

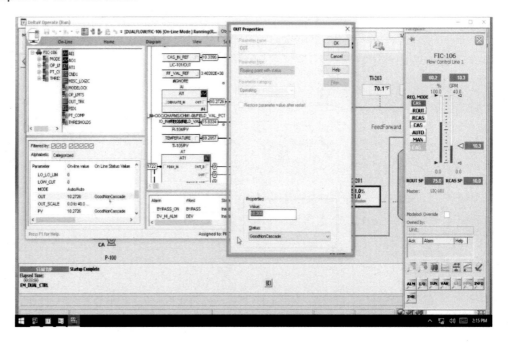

FIC-106 AI1/OUT Status=_____

Step 5.6.2

From the graphic screen, click on LIC-101 to bring up the loop faceplate.   Launch Control Studio
On-Line and record the status of the PID1 block BKCAL_IN.

Graphic 5-6-2 LIC-101 Status

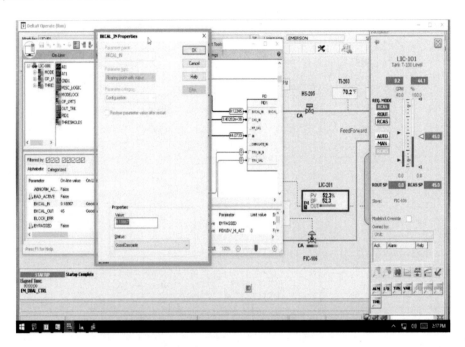

LIC-101 PID1/BKCAL_IN Status=_____

Step 5.6.3

Launch DeltaV Explorer and drill down in the Physical Network to CHM1-6 and FT-106 and launch Service Tools.

Graphic 5-6-3 DeltaV Explorer AMS Access

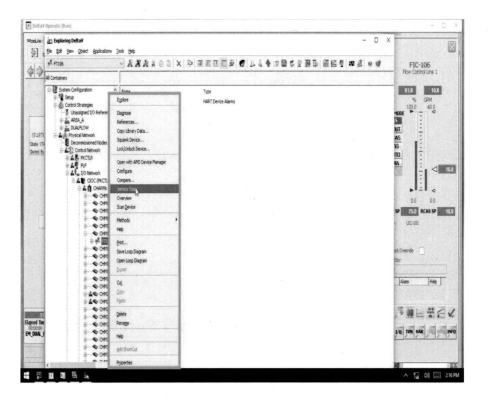

Step 5.6.4

Simulate a high saturated flow condition on FT-106 by using the Loop Test simulate function in AMS and doing an Other simulation of 21mA.

Graphic 5-6-4 AMS Loop Test

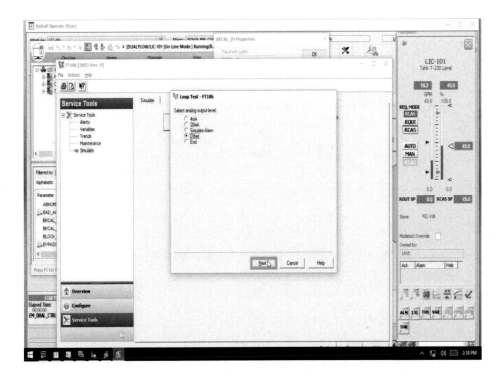

Step 5.6.5

Using Control Studio On-Line on FIC-106, record the status of the AI1 block OUT.   Does the status reflect the saturated transmitter?

FIC-106 AI1/OUT status =_____

_____

_____

Step 5.6.6

Using Control Studio On-Line on LIC-101, record the status of the PID1 block BKCAL_IN. Did the status propagate up to the Master loop?

LIC-101 PID1/BKCAL_IN status=_____

_____

_____

Step 5.6.7

Go back to AMS and remove the simulation on FIC-106.

Graphic 5-6-7 AMS Loop Test Termination

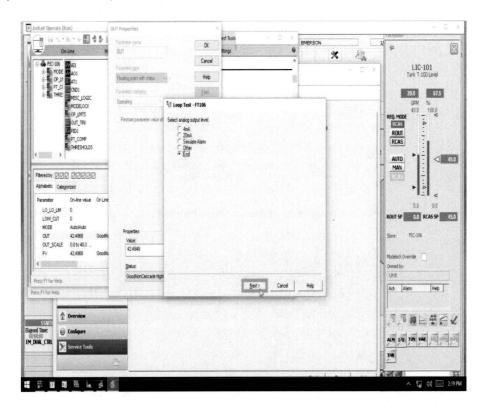

5.8   Conclusions

Every operator, process engineer and instrument technician must have a good understanding of mode to prevent quality and/or safety problems. Today's control systems are programmed to automatically react to a wide range of disturbances but there is always the possibility that one of these potential issues was overlooked. Quality and safety issues usually are the result of several abnormal conditions happening at the same time, so from time to time the operator needs to override the automatic response and put the system in manual.

Modern control systems now propagate signal status as well as values. This is typically done automatically by connecting function blocks during configuration. Prior to this functionality, the integrator had to program status propagation manually and sometimes this was overlooked in the configuration. Signal status has enabled modern control systems to react to instrumentation problems without immediate operator intervention improving quality in many industries.

For more information on Emerson training classes visit the following website:
https://www.emerson.com/en-us/automation/services-consulting/educational-services

Section 6: Control System Selection

6.1   Objectives

When the student has completed this module, the student will:

- Understand the key factors that go into selection of a control system.

- Be able to identify controller loading and free memory.

- Understand the effects of scan rate on control.

- Understand the effects of filtering on control.

6.2   Intended Audience

Instrument Technician

Process Engineer

6.3   Prerequisites

Section 2 PLP Infrastructure

Section 4 DeltaV Operate Navigation

Section 5 DeltaV Modes and Signal Propagation

6.4   Discussion

There are seven key factors that should be considered when selecting a control system for an application. These are speed, available I/O, data transfer, ease of operation, security, maintenance and cost.

6.4.1   Speed of System

The process industry typical loop speeds are relatively slow compared to a robotics system or a high-speed bottling line. Typical module scan times are 1.0 s for Flow, 2.0 s for Temperature, 1.0 s for Level and 0.2 s for Gas Pressure. A rule of thumb for module scan time should be that it is at least two times as fast as the process open loop time constant. However, normally the scan speed will be many times faster. A Distributive Control System (DCS) is ideal for the process industry but might not meet the speed requirements of other industries. There are some exceptions. For example, compressor surge control requires scan speeds of 50 ms or smaller and might require a custom controller. Not all DCS systems can handle this speed but they are improving. When determining the control system's speed, consider the I/O scan times as well.

6.4.2 Available I/O

Not all control systems support all versions of I/O. Below are some questions to consider.

- Does the system need to control/communicate with weigh systems?
- Does the system need to talk to digital transmitters like Profibus, DeviceNet, Fieldbus or ASi bus?
- Does the system need to control/communicate with camera systems?
- Does the system need to control/communicate with robotic equipment?
- Does the system need to talk to Ethernet transmitters?
- Does the system support redundancy for I/O?
- Can the I/O cards be hot swapped if they fail?
- What hazardous area I/O is supported?
- Does the system support wireless I/O?

6.4.3 Data Transfer

Not only does a control system have to control the process, but it also is the primary source of data used to troubleshoot production problems and determine production rates. The selected system must have easy methods of data extraction. In addition, many times the primary control system must communicate with smaller packaged equipment, each with their own control system. Again, the ability to do this easily should figure into the control system selection. Some question to ask are:

- Can the system easily pass data directly to other control systems? Example: Does it support Ethernet communications from a DCS to a PLC?
- Can the system easily pass data to an enterprise level continuous data historian?
- Can the system easily pass data to an asset management system for instrument calibration and maintenance?
- Can the system easily pass data to a computerized maintenance management system (CMMS)?
- Can the system easily pass data to a management execution system (MES) used for quality release and reporting?
- Can the system communicate easily with your SIS controller?

6.4.4 Ease of Operation

In many cases, there are requirements for operating consoles to be placed on the plant floor and in some rough environments. Depending on the process, touch screens may be appropriate while at other times they may cause issues. Some consoles may need to be placed in hazardous areas and need to be rated as such. Consider what operator interfaces you have in other areas of the plant. If the operator interface is the same, operators can be moved from unit to unit without re-training.

6.4.5 Security

Even before control systems were connected to the plant intranet, control system security was an important concern. It was easy to move viruses from one computer to the next with floppy drives and it is even easier with thumb drives. Some questions to ask are:

- Is the operator interface available with biometric security?
- Can the system easily support individual logons?
- Can the system easily support areas of control?
- Can the system easily support individual parameter security?
- Does the system automatically track changes in parameters and tag them to individuals?
- What remote operation capabilities are supported?
- Is the system compatible with anti-virus software?
- Can the system easily be integrated with the existing IT firewall and security infrastructure?

6.4.6 Maintenance

No control system is put in and never touched again. The system is always being changed when equipment is replaced or equipment is added. The ability to modify and maintain the system during production is extremely important to meet production goals. Some question to consider are:

- Are configuration services readily available in your area?
- Do you have controller and I/O spares already in stock?
- Can you make changes to the configuration while running?
- Can you swap out equipment while running?
- Can you do system upgrades while running?
- Can you add controllers and I/O while running?
- Are the graphics easy to modify?
- Is the system easy to configure?

6.4.7 Operating Cost

When comparing the cost of control system, always look at lifecycle costs. A system may be inexpensive to put in but may have high yearly maintenance fees for phone support and access to software revisions/patches. Some questions to ask are:

- How expensive is the system hardware?

- How expensive are the initial software licenses?

- How is the system licensed for future expansions?

- How expensive is the ongoing maintenance service contract?

- How easy is the system to configure for continuous systems?

- How easy is the system to configure for batch processes?

6.5   Workshop – Determining Controller Loading and Free Memory

Step 6.5.1

Open DeltaV Operate and from the Toolbar launch DeltaV Diagnostics.

Step 6.5.2

Expand the Control Network and click on the PKCTRL controller. Using the slider bar in the right pane find and record the Free Memory and Free Processor Time.

Free Memory = _____

Free Processor Time = _____

Graphic 6-5-2 DeltaV Control Status

6.6   Workshop – Effect of Scan Rate on Control

Step 6.6.1

From the main graphic, startup the PLP by changing the Tank Operation to STARTUP.

Step 6.6.2

From the graphic screen, click on FIC-206 to bring up the loop faceplate.   Place FIC-206 in AUTO mode and launch Process History View (PHV) from the Icon on the bottom of the faceplate.

Step 6.6.3

Change the setpoint of FIC-206 to 4.0 GPM and observe the response in PHV.

_____

_____

_____

_____

Step 6.6.4

Change the setpoint of FIC-206 to 5.0 GPM and observe the response in PHV.

_____

_____

_____

_____

Step 6.6.5

Open DeltaV Explorer from the Toolbar and expand the left pane as shown and right click on FIC-206 and select Properties.

Graphic 6-6-6 DeltaV Module Properties

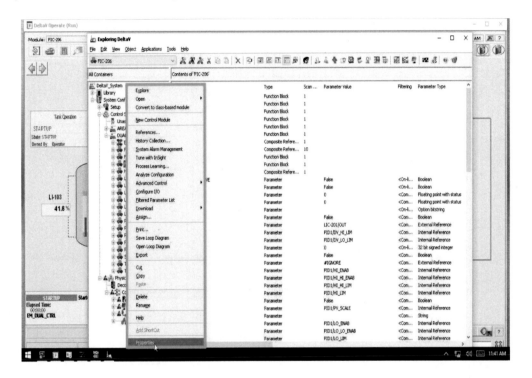

Step 6.6.6

Under the Execution tab, change the loop scan rate from 100 mS to 2 s and select OK.

Graphic 6-6-6 DeltaV Module Scan Time

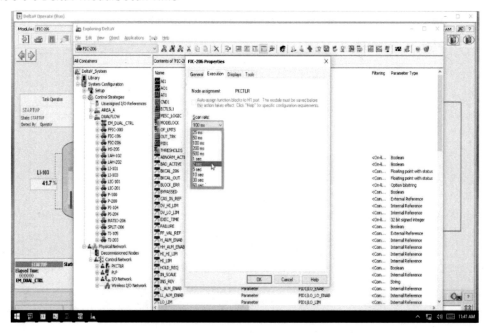

Step 6.6.7

Right click on FIC-206 again and select download control module. **Make sure to CANCEL any uploads during the download process. See screen capture below. Do not accept uploads.** The system will still try and upload nothing. Upon completion select CLOSE.

Graphic 6-6-7 DeltaV Module Download and Upload

Step 6.6.8

Repeat steps 6.6.3 and 6.6.4 and compare the results using PHV.

_____

_____

_____

_____

_____

Step 6.6.9

Return scan rate of FIC-206 to 100 mS by repeating steps 6.6.5 through 6.6.7.

6.7 Workshop – Effects of Filters on Control

Step 6.7.1

From the main graphic, startup the PLP by changing the Tank Operation to STARTUP.

Step 6.7.2

From the graphic screen, click on FIC-206 to bring up the loop faceplate.   Place FIC-206 in AUTO mode and launch Process History View (PHV) from the Icon on the bottom of the faceplate.

Step 6.7.3

Change the setpoint of FIC-206 to 4.0 GPM and observe the response in PHV.

_____

_____

_____

Step 6.7.4

Change the setpoint of FIC-206 to 5.0 GPM and observe the response in PHV.

_____

_____

_____

Step 6.7.5

From the FIC-206 Tuning Detail Faceplate, change the PV Filter TC (Time Constant) from 0.0 s to 4.0 s.

Graphic 6-7-5 DeltaV Module Filter Time

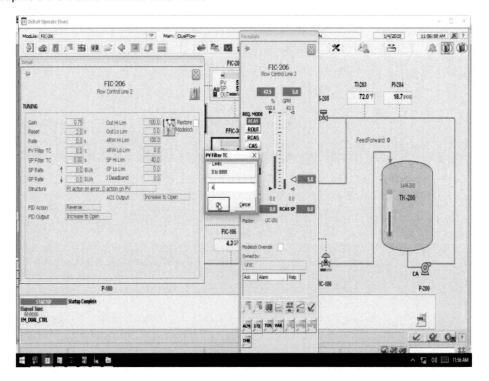

Step 6.7.6

Repeat Steps 3 and 4 and compare the results using PHV.

_____

_____

_____

_____

Step 6.7.7

Return PV Filter TC of FIC-206 back to 0.0 s.

6.8   Conclusions

With any task, there is a correct tool which allows you to accomplish the job easily and safely. A control system is no different. Selecting the wrong control system for the application can make the integration much more difficult. It could also hinder the ability to upgrade and maintain the system in the future.

For more information on Emerson training classes visit the following website:

https://www.emerson.com/en-us/automation/services-consulting/educational-services

Section 7: Level Transmitter Selection

7.1 Objectives

When the student has completed this module, the student will:

- Be able to describe the different types of continuous level transmitters available for purchase.

- Be able to explain the principles of operation of various continuous level transmitters and level switches.

- Understand how accurate level transmitters can be.

- Be able to access the configuration of a transmitter using AMS (Asset Management Software).

- Understand and describe how to strap a tank.

7.2 Intended Audience

Instrument Technician

Process Engineer

7.3 Prerequisites

Section 2 PLP Infrastructure

Section 3 PLP Instrumentation

Section 4 DeltaV Navigation

Section 5 DeltaV Modes and Signal Status

7.4 Discussion

It is critical for plant operations and safety to know the level in a vessel. If operations does not know the level in a tank, they could overfill it and cause a release. In addition, accountants and operations managers need to know the level so they can determine production rates as well as order raw materials when needed.

7.4.1   Continuous Level

7.4.1.1   Mechanical Floats

Uses Archimedes' principle. The force exerted on a float is proportional to the level in the vessel. A rough level measurement can also be a float in a site gage on the side of a tank.

- Pros

    - Inexpensive

- Cons

    - Moving parts subject to failure

    - Process contact so potential for corrosion

    - Density dependent

7.4.1.2   Nuclear

A radioactive gamma ray source is shot through a vessel to a target. The amount of radiation hitting the target is an indication of level. These are used in extremely difficult level applications and are rarely used due to the radioactive source.

- Pros

    - Measurement independent of density

    - No process contact

- Cons

    - Very expensive

    - Require a special operating permit

7.4.1.3   Differential Pressure

Static head pressure is proportional to level. The difference between two pressures is equal to the height of the liquid multiplied by the specific gravity. The capacitance between the two diaphragms on the pressure module changes when more pressure is applied. See Figure 7-4-1. The capacitance change is read via the electronics and the signal is proportional to pressure/level. LIT-103 on the PLP is a differential pressure level transmitter. If the tank is not at atmospheric, an equalization leg must be provided via an empty tube or filled capillary.

Figure 7-4-1-3 Differential Pressure Sensor

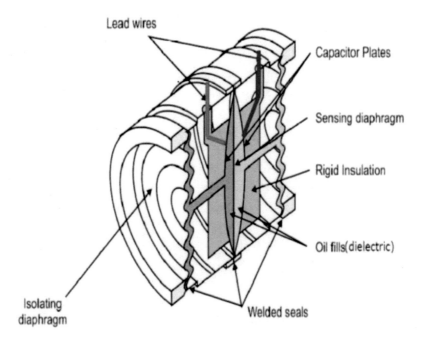

- Pros

    - Inexpensive

    - Easy to maintain and calibrate

    - Many process connections available

    - Many materials of construction available

    - Available with extended diaphragms to prevent solids build up or freezing

    - Able to suppress or elevate zero readings

    - Available with Electron Remote Sensors which can calculate the differential pressure electronically using two pressure sensors linked together eliminating thermal effects on a filled capillary

- Cons

    - Process contact required

    - Density dependent

    - Potential for freezing or solids build up at sensing port causing a false reading

    - Potential empty equalization leg could fill up with condensate causing a false reading

    - Filled capillary fluid could expand and give a false reading if used in high ambient temperature locations

–   There are limits on how much error can be compensated for. If the error
will be exceeded, on a filled capillary system, consider placing the
transmitter midway up the tank to minimize error compensation.

7.4.1.4   Radar

Microwave pulses are shot down towards the liquid and bounce off the liquid and return to the
sensor. The time it takes to return is proportional to the distance (level) from the source
divided by two. These are available as either a non-contact antenna horn or guided wave. LIT-
101 and LIT-202 on the PLP are guided wave radar transmitters.

Figure 7-4-1-4 Guided Wave Radar Operation

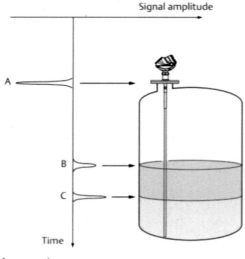

A. Reference pulse
B. Level
C. Interface level

- Pros

    –   Available in multiple materials of construction

    –   Density independent

    –   Moderate cost

    –   Antenna horn types are not in contact with process

    –   Multiple probe types available on guided wave types

    –   Guided wave type reduces chances of false echoes due to tank geometry

    –   Able to tune out false echoes due to items like agitators and ladders

    –   Able to do interface levels if process dielectric constants are different

    –   Can be used with solids

- Cons

– Subject to false readings due to foaming

– Subject to false readings due to coating of guided wave probe

– Subject to false readings due to condensate on antenna horn

– Guided wave might require a stilling well or probe support in agitated tank

– Not recommended for process fluids with extremely low dielectric constants

7.4.1.5   Ultrasonic

Sound wave pulses are shot down towards the liquid and bounce off the liquid and return to the sensor. The time it takes to return is proportional to the distance (level) from the source divided by two. With advancements in radar technology, less and less of these types of transmitters are used.

- Pros

    – Available in multiple materials of construction

    – Independent of process fluid density

    – Moderate cost

    – Antenna types are not in contact with process

    – Able to tune out false echoes due to items like agitators and ladders

    – Can be used with solids

- Cons

    – Subject to false readings due to foaming

    – Subject to false readings due to condensate on antenna horn

    – Subject to false readings if vapor space temperature, pressure or density change

7.4.1.6   Capacitance

The process fluid serves as the dielectric of a capacitor consisting of the sensor probe and the tank wall. As the tank is filled, the dielectric changes and it is proportional to the tank level.

- Pros

    – Available in multiple materials of construction

    – Density independent

    – Moderate cost

    – Able to do interface levels if process dielectrics are different and known

- Cons

    – Subject to false readings due to foaming

    – Subject to false readings due coating of probe

      – Reading dependent on dielectric constant of process; if it changes reading, the will be off.

      – Difficult to calibrate. Need to use actual process fluid or know the actual dielectric constant.

7.4.1.7 Bubblers

This is a variation of differential pressure. Air or inert gas is forced down a dip tube. The pressure required to force the gas down the dip tube is proportional to the tank level.

- Pros

      – Inexpensive

      – Many process connections available

      – Many materials of construction available

- Cons

      – Process contact required

      – Density dependent

      – Dip tube needs to be supported

      – Dip tube subject to blockage due to solids build up

      – Requires a constant flow differential regulator

7.4.2 Point Level Detection

Single point level switches are commonly used in the process industry to stop the addition of process feeds if the level in the vessel gets too high. They are considered more reliable than continuous level transmitters and thus, used in safety interlock schemes. The most common of these level switches are vibrating tuning forks. LSH-102 and LSH-202 on the PLP are vibrating fork switches. The tuning fork probe is set in motion to its natural resonant frequency via piezoelectric elements. When that vibration is disrupted by contact with a solid or liquid, a change of state is created.

The switches can be configured failsafe for low or high level switch applications and are available in multiple materials of construction including ECTFE coated for corrosive applications. They are relatively inexpensive and very reliable. The only downside is the potential for the probe to coat and give a false reading.

The PLP has two types of high level switches. LSH-102 is a common (24 VDC) discrete switch which uses non-hermetically sealed relays. Since the switch relays are not sealed, the switch must have a conduit seal if used in a classified area.

LSH-202 is a low power (24 VDC) SIL rated analog unit. The unit provides either an 8 mA or 16 mA analog signal as a change in state. This switch could be purchased with an intrinsically safe (IS) rating to meet IS wiring applications.

7.4.3 Additional Information

For more information on PLP instrumentation go to:

http://www.emersonstreamingvideo.com/pss/PLP/online/index_home_grid.html?_ga=2.1965
96328.533480057.153886.29433-271996358.1504716980

7.5 Workshop – Level Transmitter Accuracy

Step 7.5.1

Using a computer with internet access, find the Rosemount 5300 Guided Wave Radar reference manual. In the specifications section, record the published accuracy of the transmitter in water service.

Step 7.5.2

Using a computer with internet access, find the Rosemount 3051 Differential Pressure reference manual. In the specifications section, record the published accuracy of the transmitter.

Step 7.5.3

Launch DeltaV Diagnostics from the Toolbar and drill down on the left pane to CHM1-03 for LIT-101. On the right pane, use the slider bar to scroll to the channel input value.

Graphic 7-5-3 DeltaV Charm Diagnostics

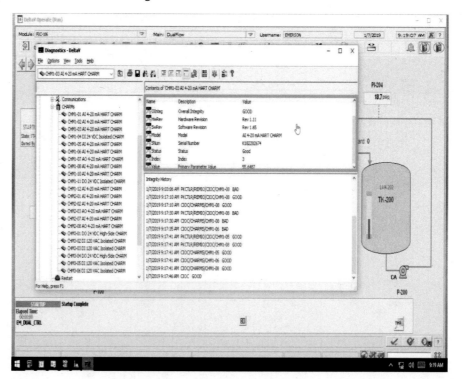

Step 7.5.4

Using a measuring cup, slowly add water to TK-100 until you see the value of CHM1-03 change. Record the amount.

LIT-101 ounces added = _____

Step 7.5.5

Repeat steps 3 and 4 for CHM1-02 (LIT-103).

LIT-103 ounces added = _____

Step 7.5.6

In DeltaV Diagnostics, return to the channel value of CHM1-02 (LIT-101) and record the amount of noise on the input.

LIT-101 Percent noise with no flow = _____

Step 7.5.7

From the main graphic in DeltaV Operate, set the complexity to Flow Control and set Tank Operation to STARTUP to get the unit running. *You may need to fill tank 1 using the FILL TANK 1 Tank Operation first prior to running STARTUP.*

Step 7.5.8

Once flows have settled down, place FIC-106 in AUTO with a setpoint of 5.0 GPM to achieve equilibrium. Allow the flows to stabilize.

Step 7.5.9

In DeltaV Diagnostics, return to the channel value of CHM1-03 (LIT-101) and record the amount of noise on the input.

LIT-101 Percent noise with flow = _____

7.6 Workshop – Level Transmitter Configuration via AMS

Step 7.6.1

Launch DeltaV Explorer from the Toolbar in DeltaV Operate.

Step 7.6.2

On the left pane, expand the Physical Network until you see LIT-103. Once there, right click on LIT-103 and select configure to launch AMS as shown below.

Graphic 7-6-2 Input AMS Access

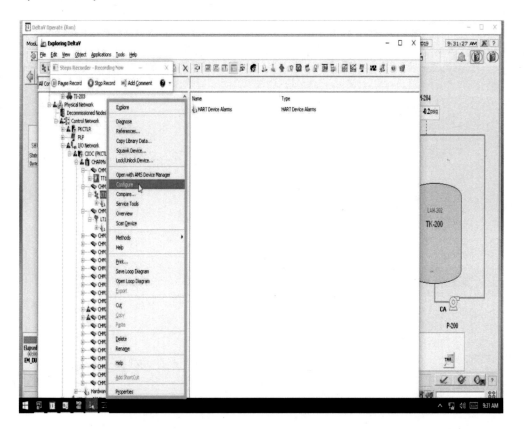

Step 7.6.3 Select Basic Setup.

Graphic 7-6-3 AMS Basic Setup Access

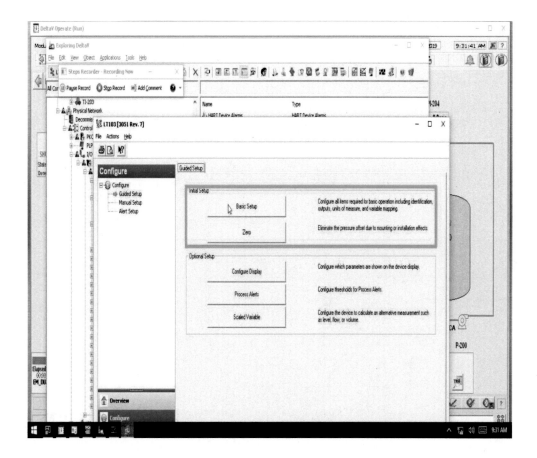

Step 7.6.4

Scroll through the Basic Configuration menu using the NEXT button without saving any changes. When you get to the PV URV (Upper Range Value) field, record the value and select CANCEL.

Graphic 7-6-4 AMS Tag Configuration

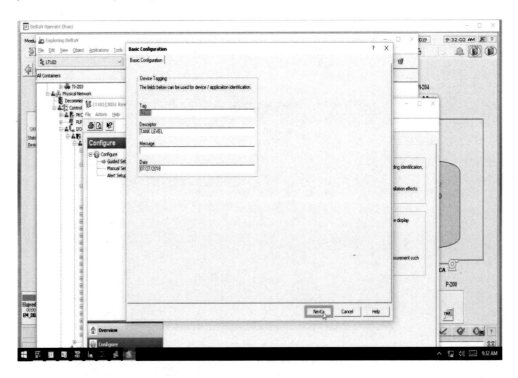

LIT-103 Upper Range Value (URV) = _____

Step 7.6.5

What would you change the URV of LIT-103 to if TK-100 was filled with ethanol with a density of 0.789 g/cm³?

LIT-103 URV for ethanol = _____

7.7 Workshop – Tank Strapping

Step 7.7.1

From the DeltaV Operate, set FIC-106 in MAN mode with an output of 50%. Place P-100 in AUTO mode and start P-100 until it stops due to low level.

Step 7.7.2

Close manual valve on outlet of TK-100.

Step 7.7.3

Drain water of TK-100 into a floor pan using the manual drain valve until it will not drain anymore. Note: TK-100 will still have some water in it.

Step 7.7.4

With a funnel, fill TK-200 manually with the water collected in the floor pan.

Step 7.7.5

Set FIC-206 in MAN mode with an output of 30%.

Step 7.7.6

Open the Interlock detail on P-200 and bypass the first 3 interlocks. Interlocks 1 and 3 are
shown bypassed below.

Graphic 7-7-6 DeltaV Interlock Bypass

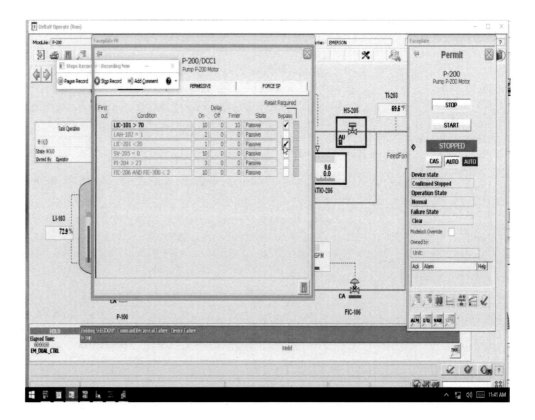

Step 7.7.7

From the Interlock detail on HS-205, bypass interlock 1.

Step 7.7.8

Using what you learned in the last workshop, launch AMS on CHM1-02 (LIT-103) and navigate to the Overview screen. Note: the current inches of water column is displayed on the left gage.

Graphic 7-7-8 AMS Process Variable Monitoring

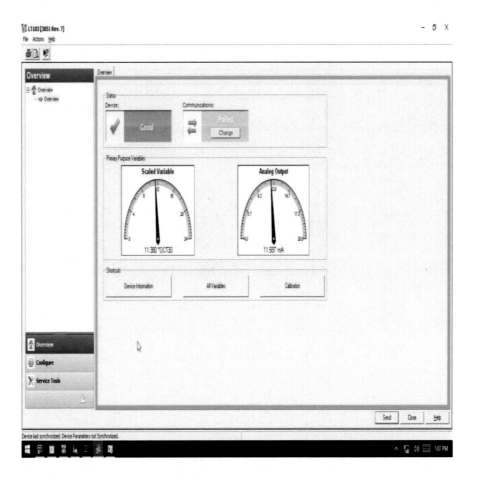

Step 7.7.9

Using what you learned in the last workshop, launch AMS via a right click on CHM1-03 (LIT-101) Process Variables. Go to the All Variables tab. Note: the current level is shown in the right-hand pane.

Graphic 7-7-9 AMS LIT-101 Process Variables

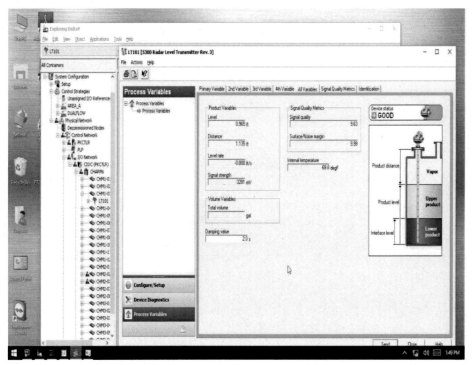

Step 7.7.10

Create an Excel spreadsheet and record the starting gallons in TK-100 using the embossed scale on the side of the tank. Also record the inches of water column of LIT-103 and distance/level on LIT-101 using AMS. In addition, record the percent level reading of LIC-101 and LIC-103 from the DeltaV Operate graphic screen.

Step 7.7.11

Place HS-205 and pump P-200 in AUTO mode. Open HS-205 and start P-200 and fill TK-100 until you reach 5 gallons as indicated on the embossed scale. Stop P-200 and close HS-205.

Step 7.7.12

Record information on Table 7-7-12 as outlined in step 7.7.10.

Table 7-7-12 Level Strapping

Actual Gallons Per Tank Embossed Scale	LIT-103 INWC	LI-103 DeltaV Percent	LIT-101 Level Feet	LIT-101 Distance Feet	LIC-101 DeltaV Percent

Step 7.7.13

Repeat steps 10 and 11 in 2 or 3 gallon increments until you reach 25 gallons in TK-100.

Step 7.7.14

From the interlock details on HS-205 and P-200 un-bypass interlocks on HS-205 and P-200.

Step 7.7.15

Plot gallons versus percent for LIC-101 and gallons vs. percent for LI-103 using figures 7-7-15A and 7-7-15B.

Figure 7-7-15A LIC-101 Gallons vs. Percent Level

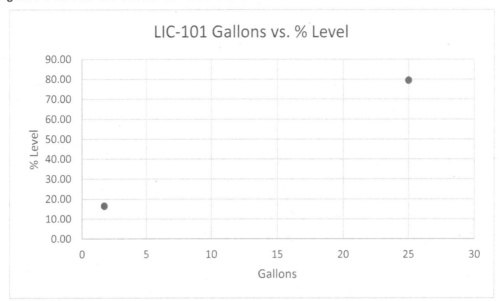

Figure 7-7-15B LI-103 Gallons vs. Percent Level

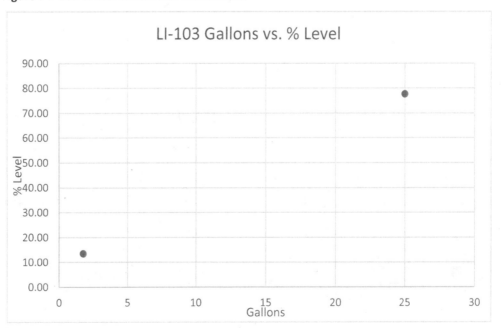

Step 7.7.16

What is unusual about the tank strapping?

_____

_____

_____

Step 7.7.18

Why is tank strapping important in an operating facility?

_____

_____

_____

7.8   Conclusions

There are numerous factors that go into the selection of a continuous level transmitter for a process vessel. Make sure you know the materials of construction that are compatible with the process fluid and the physical properties of the process fluid. Only a few of these technologies are independent of the process density, temperature and pressure. This is extremely important if you plan to use the vessel in a multi-product line.   Cost is important, but the quality and reliability of the measurement is remembered long after a potential project cost overrun.

For more information on Emerson training classes visit the following website:

https://www.emerson.com/en-us/automation/services-consulting/educational-services

Section 8: Flow Transmitter Selection

8.1 Objectives

When the student has completed this module, the student will:

- Be able to describe the different types of flow transmitters available for purchase.

- Be able to explain the principles of operation of various flow transmitters.

- Understand the impact of a poor installation on flowmeter accuracy.

- Be able to access the configuration of a flow transmitter using AMS (Asset Management Software).

8.2 Intended Audience

Instrument Technician

Process Engineer

8.3 Prerequisites

Section 2 PLP Infrastructure

Section 3 PLP Instrumentation

Section 4 DeltaV Navigation

Section 5 DeltaV Modes and Signal Status

8.4 Discussion

It is critical for plant operations to be able to accurately measure flowrates on a wide variety of gases and liquids. In a batch process, flow is used to totalize ingredient additions thereby having a direct impact on product quality. On a continuous process, flow is used to maintain production rates and to manipulate mass and energy addition rates for good composition control. Flowmeters are used extensively in the production of utilities such as steam, deionized water and plant air and are also used in custody transfer. In addition, flow is used to track plant waste discharge rates for environmental reporting compliance.

8.4.1 Flowmeter Types

8.4.1.1 Differential Pressure

Differential pressure flow measurements are based on Bernoulli's principal.   Flow rate is proportional to the square root of the differential pressure.

$$Q = K\sqrt{DP}$$

Q = flow rate

K = variable based on discharge coefficient, gas expansion factor, velocity of approach factor and bore diameter of the differential producer

DP = differential pressure at normal pressure and flow conditions

There are many types of primary differential pressure elements, these include; venturis, orifice plates, conditioning orifice plates, V-cones and annubars.

Figure 8-4-1-1 Differential Pressure Primary Elements

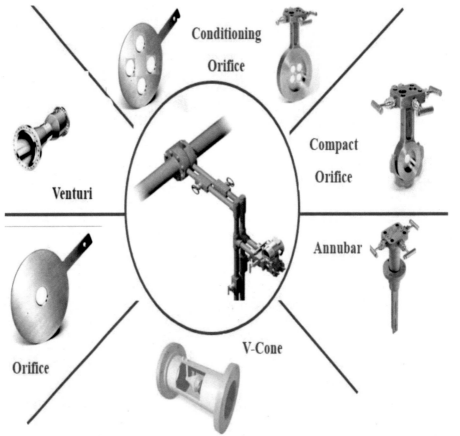

The most common and well understood primary element is the orifice plate. FIT-106 on the PLP is an orifice plate with a three-valve manifold and differential pressure transmitter. Orifice plates are used in liquid, gas and steam service and are accurate from +/- 0.5 to 3.0 % of the volumetric flow rate. The orifice diameter is based on the allowable pressure drop, pipe diameter, the desired maximum measurable flow rate and normal operating process properties (temperature, pressure and density). There are programs available from various orifice plate manufactures to properly size an orifice plate.

Figure 8-4-1-2 Example Orifice Plate Installation

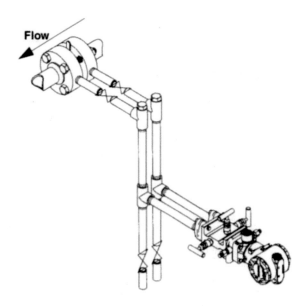

- Advantages of orifice plate installations
  - Well understood
  - Know discharge coefficient, no need for wet calibration
  - Great flexibility
  - Excellent time response
  - Bi-directional capability
- Limitations of orifice plate installations
  - Moderate turndown (from 5 : 1 to 20 : 1 turndown)
  - Impulse lines (plugging, freezing, leak points)
  - Long straight pipe run requirements (from 4 pd > 60 pd of straight pipe)
  - Many components
  - Not recommended for high solids bearing fluids
  - Significant amount of pressure drop required

8.4.1.2   Magnetic Flowmeters

Magnetic flowmeter measurements are based on Faraday's law.   This states that when a conductor passes through a magnetic field, the change in strength of a magnetic field is directly related to the velocity of the conductor.   If the cross-sectional area is known, the volumetric flow rate can be determined.   For the meter to work the process fluid must be conductive.

Graphic 8-4-1-2 Magnetic Flowmeter Operation

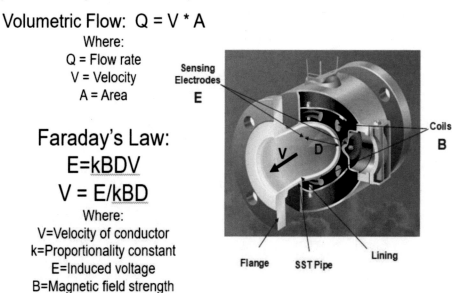

Volumetric Flow: $Q = V * A$

Where:
Q = Flow rate
V = Velocity
A = Area

Faraday's Law:
$E = kBDV$
$V = E/kBD$

Where:
V = Velocity of conductor
k = Proportionality constant
E = Induced voltage
B = Magnetic field strength
D = Length of conductor

Magnetic flowmeters are commonly used in water, acid, base and slurry services. FIT-206 on the PLP is a magnetic flowmeter.

- Advantages

    - Obstructionless, no pressure drop

    - Measures in both directions

    - High accuracy (0.15% of rate) & high rangeability (>30 : 1)

    - Available in wide variety of liner/electrode materials to handle strong acids and to strong bases to slurries

    - Process flow profile has minimal effect on measurement accuracy

    - Compatible with wide range of conductive liquids and slurries (up to 83% solids)

    - Virtually no maintenance

- Available with Smart Meter Verification including continuous verification in the background

- Limitations

  - Conductive liquids only – no gases or steam

  - Higher initial cost, especially in larger line sizes

  - 4-wire device – requires an external source of power

  - Not applicable for fluids above about 350 degrees Fahrenheit

  - 5 pipe diameters upstream and 2 pipe diameters downstream are required for the most accurate readings

8.4.1.3   Coriolis Flowmeters

Coriolis flowmeters measurements are based on the Coriolis effect. Mass flowing through a vibrating tube will create a twisting force. In the case of a dual tube Coriolis meter, as process fluid enters a sensor, the flow is split into two tubes. The flow tubes are vibrated at their natural frequency in opposition to each other by energizing a drive coil. Magnet and coil assemblies, called pick-offs, are mounted on the flow tubes. As each coil moves through the uniform magnetic field of the adjacent magnet, it creates a voltage in the form of a sine wave. If there is no flow, there is no Coriolis effect and the sine waves produced are in phase with each other. When the fluid is moving through the sensor's tubes, Coriolis forces are induced causing the flow tubes to twist in opposition to each other.   The time difference between the sine waves is measured and is called Delta-T and is proportional to the mass flow rate.

Figure 8-4-1-1 Coriolis Flowmeter Operation

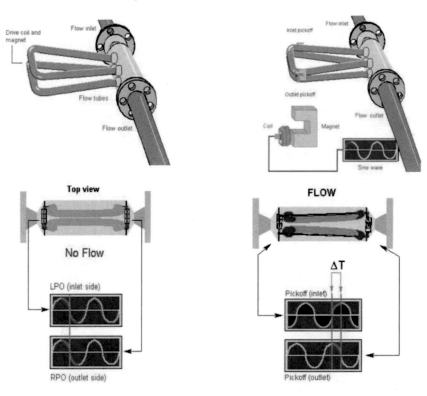

Coriolis flowmeters can be used with gases, liquids and slurries and purchased with the correct options, they can calculate concentration. They are extremely accurate (+/- 0.10% of mass flow for liquids and +/- 0.35 % of mass flow for gases/steam) and can be used for custody transfer. They have a turndown specification of greater than 20 : 1.

- Advantages

  - Direct mass flow measurement with excellent accuracy

  - Accuracy not influenced by temperature, pressure, density, viscosity or flow profile

  - One flow meter measures a wide range of fluids and flow ranges

  - No upstream or downstream piping requirements

  - Multivariable device – mass flow, volumetric flow, density and temperature in one instrument

  - Reliable due to no moving parts and requires little maintenance

  - Bidirectional

- Limitations

  - High initial cost

  - Not available in sizes larger than 16 inches

  - Relatively high pressure drop required

  - Historic challenges with entrained air

  - Typically a 4-wired device

8.4.1.4   Vortex Flowmeters

Vortex flowmeters measurements are based on the von Karman effect.   As fluid flows past a shedder bar, vortices form behind the face of the shedder bar and cause alternating differential pressure (DP) around the back of the shedder bar.   The differential pressure of the alternating vortices flexes the shedder bar and this motion is sensed by a piezoelectric sensor that converts the forces to an alternating electrical signal. The frequency of the electrical signal is proportional to the volumetric flow rate.

Figure 8-4-1-1 Vortex Flowmeter Operation

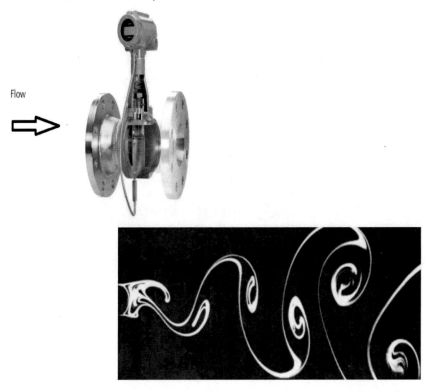

Vortex flowmeters are used on liquids, gases and steam. They have accuracy of less than +/- 1.00% of flow rate and a turndown specification of 20 : 1.   Purchased with additional pressure and temperature sensing options, they can compensate for changing process conditions and output mass flow if the process fluid properties are known.

- Advantages

    - Good accuracy and rangeability

    - Measures liquid, gas and steam

    - Low installation cost

    - Low pressure drop

    - No moving parts

    - Few connections points so minimal leaks

- Simple installation

- No pressure taps to clog

- Available with Meter Verification via Flow Simulation and Sensor Strength

- Limitations

    - Reynolds number and velocity requirements – Low Flow Cutoff

    - Often cannot measure high viscosity fluids

    - Upstream/downstream straight piping requirements (10 pipe diameters upstream, 5 pipe diameters downstream

    - Larger sizes – expensive

    - Some designs lead to potential clogging/plugging

    - Hydraulic noise and vibration – historically a problem

8.4.1.5 Ultrasonic Flowmeters

8.4.1.5.1 Doppler Ultrasonic Flowmeters

These types of ultrasonic flowmeters are based on the Doppler effect. Sound waves are shot through the process fluid and reflected back to a pickup. Due to the Doppler effect, the frequency of the sound waves changes as the velocity of the fluid increases. The frequency change is directly related to flow rate.

Figure 8-4-1-5-1 Doppler Ultrasonic Flowmeter Operation

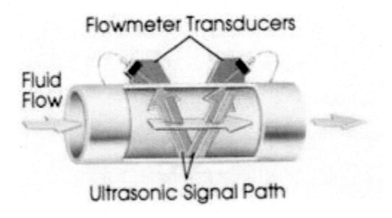

Doppler ultrasonic flow meters rely on small particles in the process fluid to create the frequency shift, so they are ideal for dirty flow streams.

8.4.1.5.2 Time-of-Flight Ultrasonic Flowmeters

Time-of-flight ultrasonic flowmeters are based on the transit time of sound pulses. Ultrasonic sound pulses are shot through the process fluid towards a transducer target. The time it takes to reach the target is directly related to the velocity of the fluid.

Figure 8-4-1-5-2 Time-of-Flight Ultrasonic Flowmeter Operation

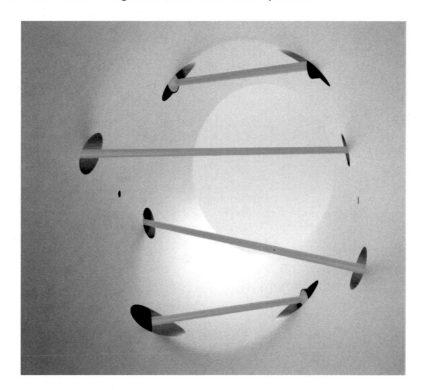

Time-of-flight ultrasonic flow meter can use a single sound path or multiple sound paths that are averaged together for the average flow reading. These flowmeters are used commonly on clean fluids and natural gas services.

8.4.1.5.3 General Ultrasonic Flowmeter Usage

Ultrasonic flowmeter's typical accuracy is 1.0 – 5.0 % of flow rate. They have a turndown specification from 20 : 1 all the way up to 50 : 1.

- Advantages
    - Clamp-on designs allow measurements to be made external to the pipe – no wetted parts
    - Newer multi-path microprocessor-based spool designs have increased accuracy in gas service
    - Very high repeatability
    - Bi-directional measurement capability
    - Obstructionless – low pressure drop

- Ideal for temporary installations or where exotic materials make a contacting flowmeter not feasible
        - No moving parts
- Limitations
        - Doppler flowmeters require dirty fluids
        - Time-of-flight flowmeter require clean fluids or gases
        - Fluid changes affect measurement
        - Proper installation and maintenance is critical (transducer alignment and dielectric gel maintenance required on transducer)
        - Long upstream and downstream straight piping requirements
        - Typically a 4-wire device

8.4.2   Additional Information

For more information on PLP instrumentation go to:

http://www.emersonstreamingvideo.com/pss/PLP/online/index_home_grid.html?_ga=2.19659632
8.533480057.153886.29433-271996358.1504716980

8.5   Workshop – Effect of Entrained Gas on Magnetic Flow Meter Operation

Step 8.5.1

Place FIC-106 in MAN mode and set the output to 50%.

Step 8.5.2

Place P-100 in AUTO mode and start P-100.

Step 8.5.3

Wait for TK-100 to drain. The system will stop P-100 automatically when it reaches its minimum level.

Step 8.5.4

Place HS-205 in AUTO mode and close it.

Step 8.5.5

Close the manual valve on the outlet of TK-100.

Step 8.5.6

Record the starting level in TK-100.

Starting level of TK-100 in percent = _____

Step 8.5.7

Set up a stop watch for 2 minutes.

Step 8.5.8

Place FIC-206 to AUTO mode with a setpoint of 5 GPM.

Step 8.5.9

As quickly as possible, open HS-205, start P-200 and start the 2 minute timer. Observe the ability of FIC-206 to control at 5 GPM.

_____

_____

_____

Step 8.5.10

At the end of 2 minutes, close HS-205 and stop P-200.

Step 8.5.11

Record the final level in TK-100 and calculate the differential level.

Final level of TK-100 in percent = _____

Differential level in percent = _____

Step 8.5.12

Repeat steps 8.5.1 through 8.5.11, but this time introduce air into the system by setting the rotameter to 20 SCFH during the test. Record the starting level, final level and differential level. Observe the ability of FIC-206 to control flow during the test.

Starting level of TK-100 in percent = _____

Final level of TK-100 in percent = _____

Differential level in percent = _____

Step 8.5.13

Was the test repeatable with air in the system? Could you tell FIC-206 was not really controlling? Why did the meter not read flow correctly?

_____

_____

_____

Step 8.5.14

Open the main graphic and start the flow on the PLP using Complexity: Flow Control.

Graphic 8-5-1 PLP Startup

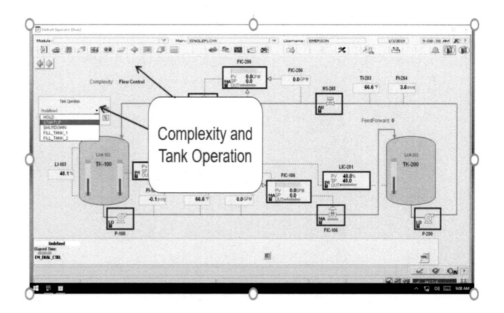

Step 8.5.15

Launch DeltaV Explorer from the main graphic Toolbar and drill down in the Physical Network to CHM2-02 for FIT-206 and launch AMS by right clicking and select Service Tools.

Graphic 8-5-2 AMS Access of FT-206

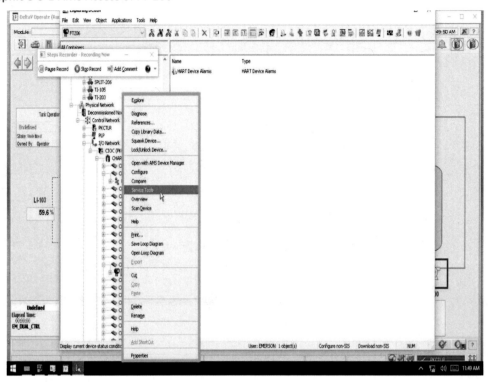

Step 8.5.16

Once in AMS, open Service Tools and select Variables in the left-hand pane. Go to the Process
Diagnostics tab and record the5 Hz Signal to noise ratio value and the 37 Hz Signal to noise ratio
under normal conditions.

Graphic 8-6-3 FT-206 Magnet Flowmeter AMS Diagnostic

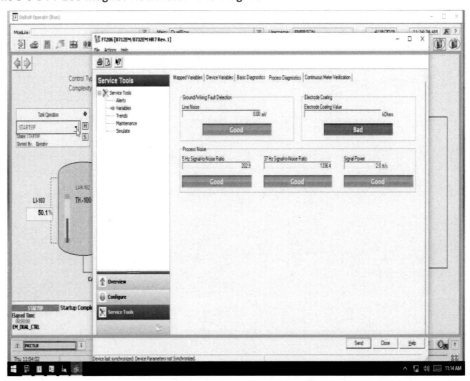

Normal 5 Hz Signal to Noise Ratio = _____

Normal 37 Hz Signal to Noise Ratio = _____

Step 8.5.17

Repeat steps 8.5.16, but this time introduce air into the system by setting the rotameter to 20 SCFH during the test. Why did the signal to noise ratio diagnostic values change? The meter has the ability to operate at 37 Hz, would that help the meter deal with entrained gases?

5 Hz Signal to Noise Ratio with Entrained Gas =_____

37 Hz Signal to Noise Ratio with Entrained Gas = _____

_____

_____

_____

8.6 Workshop – Orifice Plate Differential Pressure Flow

Step 8.6.1

Launch DeltaV Explorer from the main graphic Toolbar and drill down in the Physical Network to CHM1-06 for FIT-106 and launch AMS by right clicking and select Overview.

Graphic 8-6-1 AMS Access of FT-106

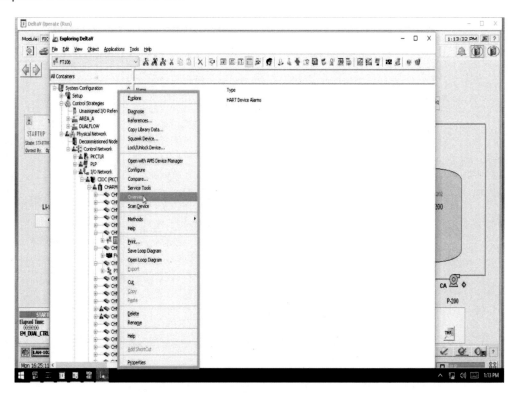

Step 8.6.2

Once in AMS, select Configure in the left pane and click on Scaled Variable to begin the Scaled Variable menu. Continue through the menu by selecting the Next button and not making any changes. When you get to the last menu item, select Cancel. Determine if the Square Root function is enabled and the inH2O to GPM upper scaling values.

Graphic 8-6-2 FT-106 Scaled Variable

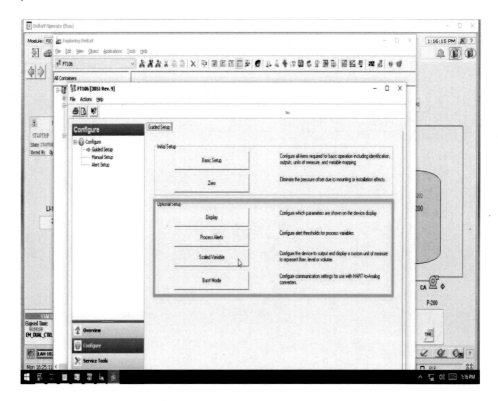

Square Root Enabled? =._____

Upper inH20 Scaled Value = _____

Upper GPM Scaled Value = _____

Step 8.6.3

Open the main graphic and start the flow on the PLP using Complexity: Flow Control.

Graphic 8-6-3 PLP Startup

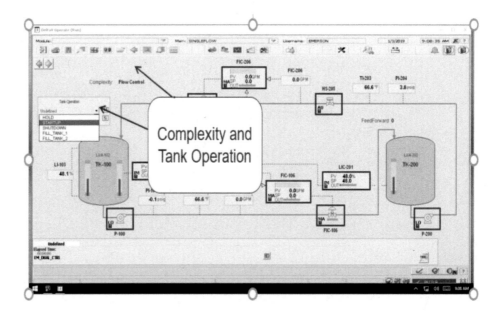

Step 8.6.4

Allow the FIC-106 to stabilize at about 5 GPM.

Step 8.6.5

In AMS under Overview, select All Variables. Record the flow in GPM and the pressure in inH2O.

Flow Rate = _____

Pressure = _____

Step 8.6.6

On FIT-106, open the green equalization valve on the three valve manifold. Record the flow in GPM and the pressure in inH2O. Why did the pressure go down? Why didn't the pressure go to zero?

Flow Rate = _____

Pressure = _____

_____

_____

_____

Step 8.6.7

Close the green equalization valve.

Step 8.6.8

From an internet accessible computer, go to the following website and download an orifice plate calculator.

https://www.spartancontrols.com/documentation/document-library/measurement-instrumentation/daniel/software/daniel-orifice-flow-calculator/

Step 8.6.9

Calculate the orifice diameter for water service with a 0 – 30 GPM transmitter using 300 INWC pressure drop in a 1 inch pipe.

Orifice diameter = _____ inches

8.7 Workshop – Effect of Entrained Gas on Coriolis Flow Meter Operation

Step 8.7.1

Open the main graphic and start the flow on the PLP using Complexity: Flow Control.

Graphic 8-7-1 PLP Startup

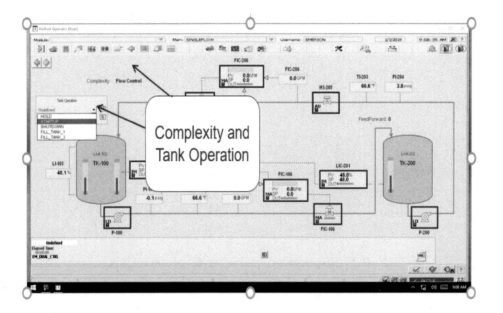

Step 8.7.2

Launch DeltaV Explorer from the main graphic Toolbar and drill down in the Physical Network to CHM2-02 for FIT-206 and launch AMS by right clicking and select Service Tools.

Graphic 8-7-2 AMS Access of FT-206

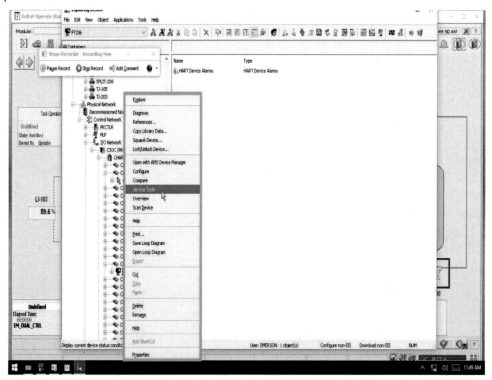

Step 8.7.3

In AMS, click on Variables in the left pane and open the Process tab.   Record the volumetric flow rate, mass flow rate and density.

Volumetric Flow Rate = _____

Mass Flow Rate = _____

Density = _____

Graphic 8-7-3 Normal Coriolis Readings

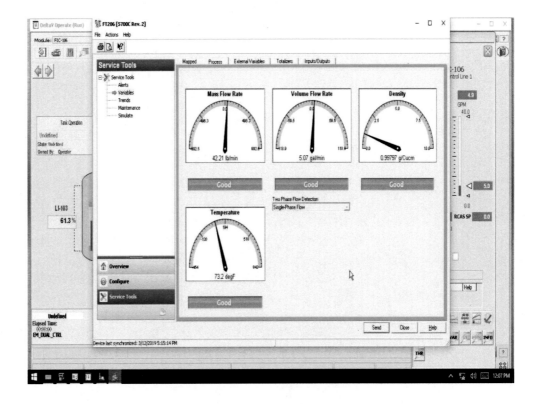

Step 8.7.4

In AMS, click on Maintenance in the left pane and open the Diagnostic Variable tab.   Record the drive gain.

Drive gain = _____

Graphic 8-7-4 Normal Coriolis Drive Gain

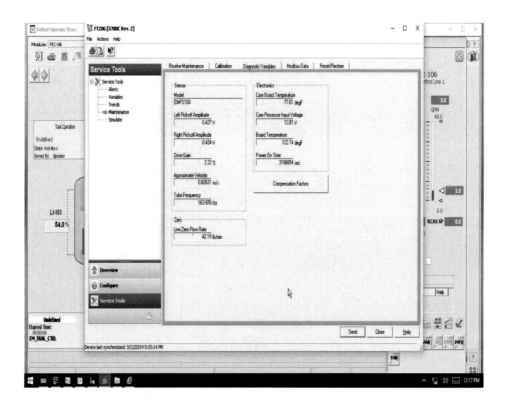

Step 8.7.5

Repeat steps 8.7.3 and 8.7.4, but this time introduce air into the system by setting the rotameter to 20 SCFH during the test. Record the volumetric flow rate, mass flow rate, density and drive gain. What is the significance of the drive gain?

Volumetric Flow Rate = _____

Mass Flow Rate = _____

Density = _____

Drive gain = _____

_____

_____

8.8   Workshop – Zeroing a Coriolis Meter

Step 8.8.1

Open the main graphic and start the flow on the PLP using Complexity: Flow Control.

Graphic 8-8-1 PLP Startup

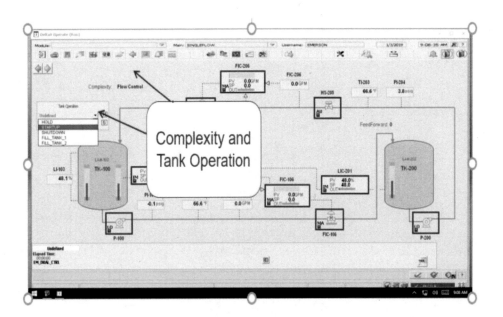

Step 8.8.2

In AMS under Service Tools, click on Maintenance in the left pane and open the Calibration tab. Click on the Zero Calibration button.

Graphic 8-8-2 AMS Coriolis Flowmeter Zero

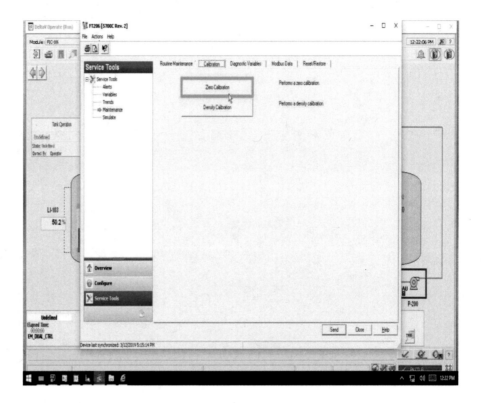

Step 8.8.3

To properly zero a Coriolis meter, there meter needs to be full and the meter needs to be installed in the pipeline.   Record the Zero value. Try to perform a zero calibration with the PLP running use a 20 second zero time.   What happens to the Zero value?   Shutdown the PLP and record the volumetric flow rate with no flow.

Zero value before Zero Procedure = _____

Zero value after Zero Procedure = _____

No Flow Volumetric Flow Rate = _____

_____

_____

Step 8.8.4

With the PLP shutdown, restore the factory zero by clicking on the Restore the Factory Zero button. Record the Zero value.

Factory Zero value = _____

Step 8.8.5

With the PLP shutdown, perform a 20 second zero calibration by clicking on the button Perform a Zero Calibration button. Record the Zero value. Why is the new zero value different from the factory zero?

Zero value = _____

_____

_____

8.9 Workshop – Effect of Vortex Flowmeter Low Flow Cutoff Setting

Step 8.9.1

Open the main graphic and start the flow on the PLP using Complexity: Flow Control.

Graphic 8-9-1 PLP Startup

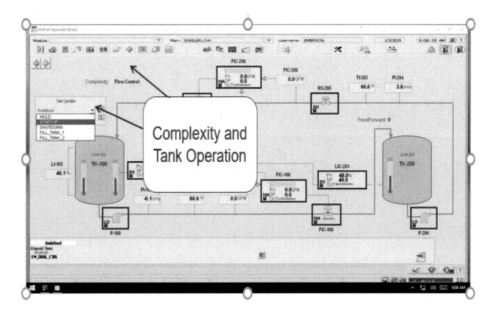

Step 8.9.2

Launch DeltaV Explorer from the main graphic Toolbar and drill down in the Physical Network to CHM2-07 for FIT-300 and launch AMS by right clicking and select Configure.

Graphic 8-9-2 AMS Access of FT-300

Step 8.9.3

In AMS under Configure, click on Manual Setup in the left pane and open the Signal Processing tab. Record the Low Flow Cutoff in Engineering Units, Low Flow Cutoff and the Recommended Minimum Low Flow Cutoff.

Low Flow Cutoff in Engineering Units = _____

Low Flow Cutoff Frequency = _____

Recommended Low Flow Cutoff = _____

Graphic 8-9-2 Vortex Flowmeter AMS Signal Processing Menu

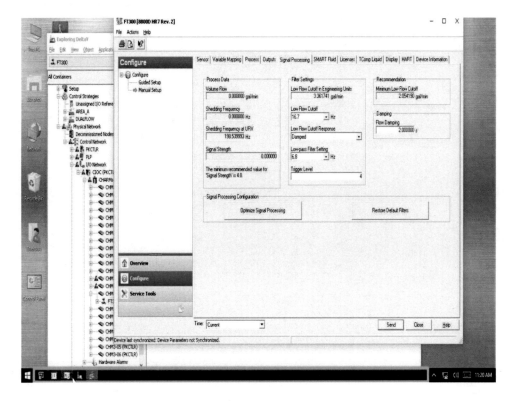

Step 8.9.4

Optimize the signal processing using a density of 40 lb/Cuft by clicking the Optimize Signal Processing button. Did the low flow cutoff values change?

_____

_____

_____

Step 8.9.5

Place FFIC-300 in MAN mode and slowly decrease the output in 1% increments until the flow reading on the vortex meter drops out.   How low could you get the output before the vortex meter output went to 4 mA?

Lowest measurable flow rate = _____

Percent valve open at lowest measurable flow rate = _____

Step 8.9.6

Decrease the Low Flow Cutoff frequency from its current value of 16.7 Hz until the Low Flow Cutoff in Engineering Units rate approaches the recommended Minimum Low Flow Cutoff flow rate. Send it down to the Vortex meter using the Send button on the lower right corner of the AMS screen. Record the final Low Flow Cutoff frequency and the Low Flow Cutoff in Engineering units.

Low Flow Cutoff in Engineering Units = _____

Low Flow Cutoff Frequency= _____

Step 8.9.7

Repeat step 8.9.5.   Was the lowest measurable flow rate on FFIC-300 lower?

Lowest measurable flow rate = _____

Percent valve open at lowest measurable flow rate = _____

_____

_____

_____

Step 8.9.8

Restore the Low Flow Cutoff frequency to 16.7 Hz and Send it to the device.

8.10   Conclusions

There are numerous factors that go into the selection of a flow transmitter for a specific application. Make sure you know the materials of construction that are compatible with the process fluid and the physical properties of the process fluid. Also, know the accuracy and range requirements of the measurement. Flowmeters are only good over a specific range and the accuracy drops off dramatically at the lower end. Only a few of these technologies are independent of the process density, temperature and pressure. This is extremely important if you plan to use the meter to measure multiple process streams. Cost is important; however, the quality and reliability of the measurement is remembered long after a potential project cost overrun.

For more information on Emerson training classes visit the following website:

https://www.emerson.com/en-us/automation/services-consulting/educational-services

Section 9: Pressure Transmitter Selection

9.1 Objectives

When the student has completed this module, the student will:

- Be able to describe the different types of pressure transmitters/transducers available for purchase.

- Be able to explain the principles of operation of various pressure transmitters/transducers.

- Be able to zero a pressure transmitter.

- Be able to access the configuration of a pressure transmitter using AMS (Asset Management Software).

9.2 Intended Audience

Instrument Technician

Process Engineer

9.3 Prerequisites

Section 2 PLP Infrastructure

Section 3 PLP Instrumentation

Section 4 DeltaV Navigation

Section 5 DeltaV Modes and Signal Status

9.4 Discussion

It is critical for plant operations to be able to accurately measure pressure on a wide variety of gases and liquids. Pressure is the primary measurement on utility systems such as steam, instrument air and nitrogen. Reliable pressure measurements are essential for the safe operation of a plant as well. In many cases, pressure is an early indication of a possible release. From a maintenance standpoint, it confirms the system is not under pressure prior to line entry and is a key indication that equipment is functioning within specifications.

9.4.1 Pressure Transducers and Transmitters

9.4.1.1 Capacitance Pressure Transmitters

The capacitance between the two diaphragms on the pressure module changes when more pressure is applied. See Figure 9-4-1-1. The capacitance change is read via the electronics and the signal is proportional to pressure. PIT-104 and PIT-204 on the PLP are capacitance-based pressure transmitters. These transmitters are capable of gage pressure, absolute vacuum and differential pressure measurements and are the workhorse of the process industry.

Figure 9-4-1-1 Pressure Sensor

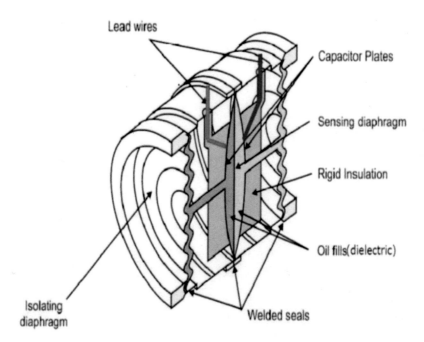

- Pros

    - Inexpensive and reliable

    - Available with remote diaphragms with filled capillaries to prevent freezing

    - Can be used with gases or liquids

    - Easy to maintain and calibrate

    - Many process connections available

    - Many materials of construction available

    -

- Cons

    - Impulse lines can freeze and may require heat tracing if mounted away from the process

– Filled capillary fluid could expand and give a false reading if used in high ambient temperature locations

9.4.1.2   Piezoelectric Pressure Transducers

When pressure is applied to certain types of crystals, an electrical potential is produced within the crystals causing an electrical current to flow. This current is proportional to the force applied. These transducers are available with small quartz crystals and ceramic crystals allowing use in small footprint applications. They are used in blast and explosion monitoring such as ballistics.

- Pros

    – Inexpensive and reliable

    – Small foot print

    – High speed response

- Cons

    – Not for vacuum applications

    – Only a dynamic measurement

9.4.1.3   Strain Gage Transducers

Strain gage transducers are commonly used in weighing applications in many production facilities. An example is a common load cell. The operating principle of a resistance-type strain gage is based on a change of resistance when a conductor is compressed or elongated. Resistance is a function of resistivity, length and cross-sectional area. If pressure is applied to a load cell, the resistance of the load cell is changed, and that resistance change is proportional pressure.

- Pros

    – Very accurate

    – High speed response

- Cons

    – Zero drift

9.4.1.4   Additional Information

For more information on PLP instrumentation go to:

http://www.emersonstreamingvideo.com/pss/PLP/online/index_home_grid.html?_ga=2.19659632
8.533480057.153886.29433-271996358.1504716980

9.5 Workshop – Zeroing a Pressure Transmitter

Step 9.5.1

Open DeltaV Operate via the FlexLock Banner.

Step 9.5.2

From the graphic screen, set the Complexity to Flow Control and Tank Operation to STARTUP.   PLP
will startup. Allow FIC-106 and FIC-206 to stabilize at around 5.0 GPM. *You may need to run Fill
Tank 1 first to get TK-100 above 55%.*

Graphic 9-5-1 PLP Startup

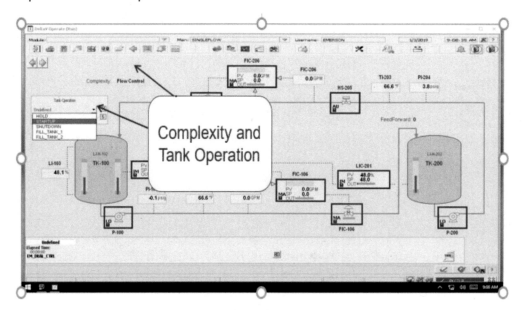

Step 9.5.3

Record the pressure reading of PI-204 from DeltaV.

Pressure PI-204 = _____

Close the blue isolation valve on the PIT-204 pressure transmitter isolation manifold.

Step 9.5.4

Close the red vent valve on PIT-204 pressure transmitter isolation manifold if not already closed.

Step 9.5.5

Remove the vent plug on the PIT-204 pressure transmitter isolation manifold and slowly open the
red vent valve on the manifold.

Step 9.5.6

Record the pressure reading on PI-204 from DeltaV.

Pressure PI-204 = _____

Step 9.5.7

From the main graphic Toolbar, launch DeltaV Explorer.

Step 9.5.8

In the left pane, under the Physical Network, drill down to CHM1-08 for PIT-204. Right click on PIT-204 and select Overview to launch AMS.

Graphic 9-5-8 DeltaV AMS Initiation

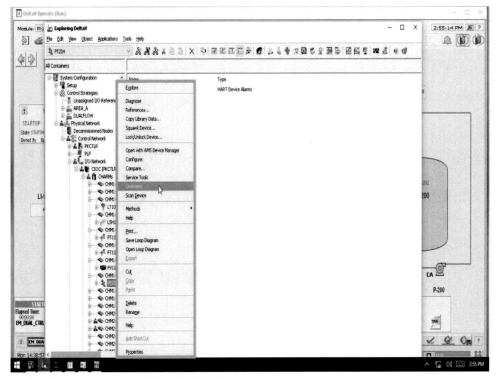

Step 9.5.9

In AMS, go to Configure/Guided Setup and click the Zero button to begin the zero procedure. Keep clicking the Next button until the procedure completes. Make sure the transmitter is vented to atmosphere before executing the zero procedure.

Graphic 9-5-9 Transmitter Zero

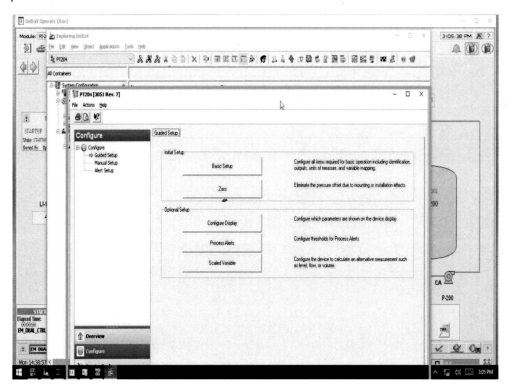

Step 9.5.10

Record the pressure reading on PI-204 from DeltaV. Did the transmitter change to zero PSIG?

Pressure PI-204 = _____

Step 9.5.11

Replace the vent plug and record the pressure reading on PI-204 from DeltaV. Why did the pressure go up?

Pressure PI-204 = _____

Step 9.5.12

Close the red vent valve and record the pressure reading on PI-204 from DeltaV. Why did the pressure go up?

Pressure PI-204 = _____

_____

_____

Step 9.5.13

Open the blue isolation valve on the manifold and record the pressure reading on PI-204 from DeltaV. Did the pressure return to the value in step 9.5.3?

Pressure PI-204 = _____

_____

_____

Step 9.5.14

Due to space constraints PIT-204 is installed above the sensing point in liquid service. Why could this be a problem in extremely sensitive pressure measurement applications?

_____

_____

_____

9.6   Conclusions

There are numerous factors that go into the selection of a pressure transmitter for a specific application. Make sure you know the materials of construction that are compatible with the process fluid and the physical properties of the process fluid. Also, know the accuracy and the range requirements of the measurement. Pressure transmitters are only good over a specific range. Transmitter placement is critical for an accurate measurement due to solids build up or compressible gases being trapped between the sensor and liquid. Consider instrument accessibility and always install isolation valves and vent valves so the transmitter can be zeroed and calibrated in line.

For more information on Emerson training classes visit the following website:

https://www.emerson.com/en-us/automation/services-consulting/educational-services

Section 10: Control Valve Selection

10.1 Objectives

When the student has completed this module, the student will:

- Be able to describe some of the different types of final control elements.

- Be able to explain the principles of operation of various control valves.

- Be able to describe the different trim packages available on globe valves.

- Understand the importance of fail position on valves.

- Understand what a valve positioner does.

- Be able to access the configuration of a control valve positioner using AMS (Asset Management Software).

10.2 Intended Audience

Instrument Technician

Process Engineer

10.3 Prerequisites

Section 2 PLP Infrastructure

Section 3 PLP Instrumentation

Section 4 DeltaV Navigation

Section 5 DeltaV Modes and Signal Status

10.4 Discussion

Process feedback loops consist of three components: the measurement, the controller and the final control element. The final control element is the way in which the controller affects the process. There are numerous final control elements on the market, some for discrete control and some for analog control.

10.4.1 Discrete Control

Discrete control is used throughout the process industry. For example, if the level in a sump pit gets too high, the control system will turn on the pump and open a valve. Another example would be a thermostat which turns on steam to a steam coil when the temperature drops a couple of degrees below the setpoint.

One of the most common final control elements used in discrete control is the On/Off valve. There are many types of On/Off valves on the market, but all interact with the control system via an actuator. An actuator is used to supply the force required to open or close a valve. Control systems are only able to drive small loads, and in many cases, there is not enough energy available to open or close large valves. To overcome this, the control system opens a small solenoid valve which releases high pressure air to trigger an actuator. Below are some of the types of On/Off valves available.

- Ball valves

- Plug valves

- Diaphragm valves (Example: HS-205 on the PLP)

- Gate valves

From time to time, utilities such as instrument air and electricity go down due to equipment failures. Therefore, it is critical to safety that each On/Off valve have an actuator with a predesignated fail position. This fail position should be identified on the P&ID and be determined by the process engineer to prevent an unsafe excursion. The fail action is achieved by springs that are compressed when the actuator is in the non-failed state. Most valves fail in the closed position, but in some cases like emergency cooling, the valve may need to fail in the open position. Valves with piston actuators can also fail in the last position. Fail last position valves have no springs installed in the actuator and require two solenoids or a trip valve.

10.4.2 Analog Control

An analog controller manipulates the final control element in small increments to maintain setpoint. The more responsive and faster the final control element responds to controller output moves, the more precise the control can be. Examples of final control elements are:

- Variable Frequency Drives (VFD)

  - VFDs can increase and decrease the speed of a motor

  - They contribute minimal dead time to the control loop

- Mechanical Dampers

  - Used to control air flow

  - Loose linkages and friction losses can contribute to loop dead time

- Heaters

  - Used with Silicon Controlled Rectifiers (SCR) to provide precise temperature control by varying the frequency a heater is cycled on and off

- Control Valves

  - Most common analog final control element

  - Can be installed with positioners to overcome valve hysteresis and stem friction

10.4.3   Control Valves

There are many types of control valves on the market. They typically have a spring and diaphragm air actuator and are installed with a transducer that converts the control system electrical output signal to a pneumatic signal. Control valves actuators are required to provide the energy to move the valve to the correct position. The forces that must be overcome to properly position the valve are process unbalance pressure, packing friction, seat load and other friction such as valve plug seals. All these forces combine and result in causing the valve and actuator assembly to have deadband and hysteresis. To overcome valve deadband and hysteresis some control valves are installed with positioners. Positioners have feedback sensors that sense the actual valve stem position and supply the required energy needed to move the valve to the desired position coming from the control system. Remember, for a precise control, the final element needs to be responsive to the smallest of output changes and perform those control moves quickly and without overshoot. Examples of control valves are:

- Butterfly Valves

    - Used in large piping applications due to cost

    - Above 60% open stability issues can be encountered with none-FishTail disks

- V-Ball Valves

    - Used in slurry service

    - Used in many steam and chemical applications where pressure drops aren't excessive

- Globe Valves

    - Most common control valves in small piping applications (Example of globe valves: FV-106 and FV-206 on the PLP)

    - Available with many trim sizes and flow characteristic types

        - Equal Percentage

        - Linear

        - Fast Opening

Figure 10-4-3 Globe Control Valve Response

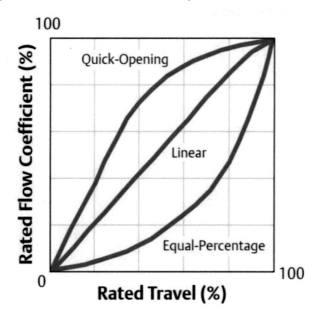

For optimal PID control, the response to output changes must be linear. In other words, a 10% change in valve position when the valve moves from 10% to 20% should produce the same response in the measured variable as if the valve was at 70% open and moved to 80%. In a case where there is constant pressure drop across a control valve, linear trim will produce a linear response. However, in most process applications, the pressure drop across the valve decreases as the flow rate increases; thus, in most centrifugal pump applications equal percentage trim is recommended to produce a linear response. This is due to the fact that the inlet pressure to the valve decreases as the flow rate increases as a result of the "pump curve".

10.4.4   Additional Information

For more information on PLP instrumentation go to:

http://www.emersonstreamingvideo.com/pss/PLP/online/index_home_grid.html?_ga=2.196596328.533480057.153886.29433-271996358.1504716980

10.5   Workshop – Control Valve Failure

Step 10.5.1

Click on FIC-206 on the main graphic and place it in MAN mode with an output of 50%. Verify FV-206 is at 50% using the site gage on the valve itself.

Step 10.5.2

Close the manual valve supplying air to the PLP and wait 2 minutes. What happened to the control valve position on the site gage? What is the fail position of FV-206?

_____

_____

_____

Step 10.5.3

Open the manual air valve to PLP.

10.6 Workshop – Positioner Functionality

Step 10.6.1

From the main graphic, click on FIC-206 and place in MAN mode with an output of 0%.

Step 10.6.2

From the main graphic Toolbar, launch DeltaV Explorer. Drill down to CHM02-03 and right click and select Overview to launch AMS.

Graphic 10-6-2 Valve AMS Access

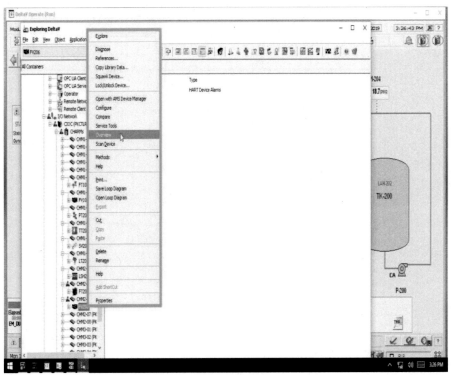

Step 10.6.3

Place a metal obstruction between the flat travel indicating disk on FV-206 and the green actuator body. **A large socket from a ratchet set works well, but whatever obstruction is used, it should be shielded from people performing the test just in case it becomes dislodged.** Slowly increase the output of FIC-206 in 1% increments until the obstruction just starts to become wedged between the actuator body and the indicating disk. Record the Setpoint, the pressure going to the valve (Pressure A) and the Travel (actual valve position).

Setpoint % = _____

Pressure A in PSIG = _____

Travel % = _____

Graphic 10-6-3 Position AMS Readings

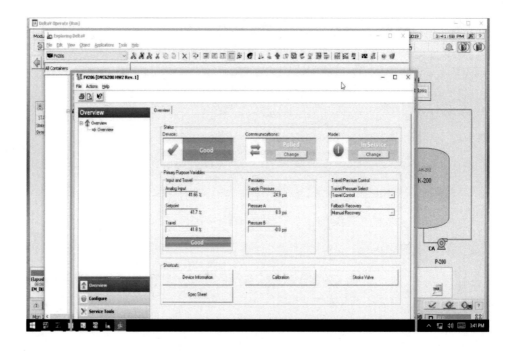

Step 10.6.4

From the main graphic click on FIC-206 and set the output to 0%.

Step 10.6.5

Remove the obstruction and set the output to the setpoint value in step 10.6.3. Record the Setpoint, the pressure going to the valve (Pressure A) and the Travel (actual valve position). Why is the pressure going to the valve (Pressure A) significantly different?

Setpoint % = _____

Pressure A in PSIG = _____

Travel % = _____

_____

_____

_____

10.7   Workshop – Control Valve Response

Step 10.7.1

Click on P-200 on the main graphic and select the interlock detail for P-200. Bypass the low flow interlock condition number 6 (FIC-206 < 2 GPM).

Step 10.7.2

From the graphic screen, set the Complexity to Flow Control and Tank Operation to STARTUP. PLP will startup. Allow FIC-106 and FIC-206 to stabilize at around 5.0 GPM. *You may need to run Fill Tank 1 first to get TK-100 above 55%.*

Graphic 10-7-2 PLP Startup

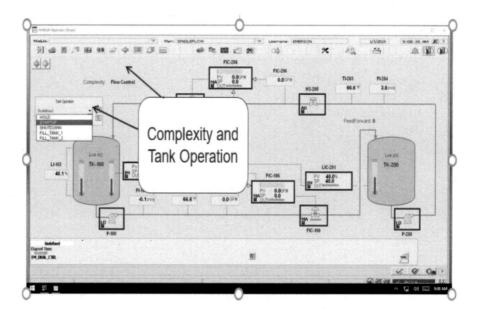

Step 10.7.3

Click on FIC-206 and place in MAN mode with an output of 5%. Record the pressure of PI-204 and the flow rate on FIC-206 in Table 10.7.3. Repeat this step in 5% increments. *Note: you may need to fill TK-200 back up from time to time using the FILL TANK 2 procedure.*

Table 10.7.3 Pressure and Flow Rates and Cv for FIC-206 at Different Valve Percentages

FV-206 % Open	PI-204 Pressure in PSIG P1	Pressure Drop across valve P1 -R assuming constant line losses	Flow of FIC-206 in GPM Q	Adjustment Factor F	Calculated $Cv = \frac{F\,Q}{\sqrt{P1-R}}$
0	-	-	0	-	0
5					
10					
15					
20					
25					
30					
35					
40					
45					
50					
55					
60					
65					
70					
75					
80					
85					
90					
95					
100					9.5

Calculate the line losses R and the adjustment factor F based on the following information:

- Max Cv is 9.5 when the valve is fully open because it is marked on the valve.

- TK-100 is at zero PSI since it is open to atmosphere.

- The pressure drop across the valve in the full open position is at its minimum. Based on the data collected, assume a pressure drop across the valve of 0.3 PSI. The line losses can be estimated as pressure PI-204 at 100% minus the pressure drop across the valve at the full open position.

R = P1 at 100% - 0.3

$$F = \frac{Cv\sqrt{P1-R}}{Q}$$

R = _____PSI

F = _____

Calculate Cv using formula

$$Cv = \frac{F\,Q}{\sqrt{P1-R}}$$

Q = Flow rate

P1 = Inlet Pressure PI-204

P2= Pressure on other side of control valve

F = Adjustment factor based on valve design, pipe geometry and Reynolds number

Plot Flow versus percent open and Cv versus percent open in EXCEL or by hand. What type of trim is FV-206?

Figure 10-7-3-1 FIC-206 Flow Versus FV-206 % Open

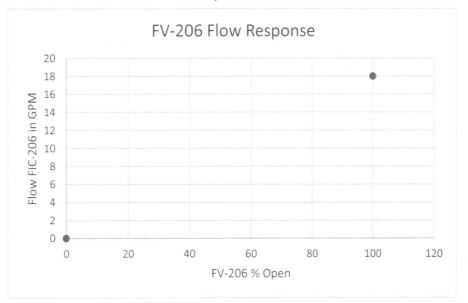

Figure 10-7-3-2 FV-206 % Open Versus Cv

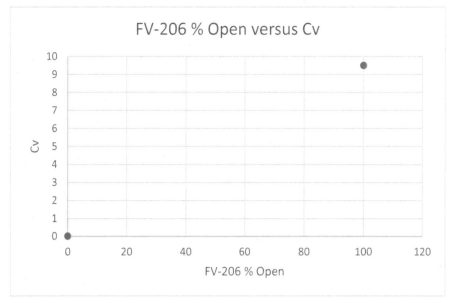

FV-206 Valve Trim = _____

Step 10.7.4

Reinstate interlock on P-200.

10.8   Conclusions

There are numerous factors that go into the selection of a control valve for a specific application. Make sure you know the materials of construction that are compatible with the process fluid and the physical properties of the process fluid. The correct fail position is critical for predictable and safe operation if there is a loss of utilities. A positioner will compensate for valve deadband and hysteresis, increased stem friction due to coating and high friction packing and it is usually worth the added expense for precise control. Finally, be aware that the valve installed characteristic is different than the inherent characteristic and that the installed characteristic should be close to linear for optimal PID control.

For more information on Emerson training classes visit the following website:

https://www.emerson.com/en-us/automation/services-consulting/educational-services

Section 11: I/O Types

11.1 Objectives

When the student has completed this module, the student will:

- Be able to describe all of the types of I/O available on the market.

- Be able to locate the I/O channels associated with a DeltaV module.

- Be able to describe the advantages of electronic marshalling.

11.2 Intended Audience

Instrument Technician

11.3 Prerequisites

Section 2 PLP Infrastructure

Section 3 PLP Instrumentation

Section 4 DeltaV Navigation

Section 5 DeltaV Modes and Signal Status

11.4 Discussion

I/O (Inputs and Outputs) is the way the control system interacts with the process. There are numerous types on the market and each vendor has their own customized I/O offerings. In some cases, there is third party I/O that integrates well with a control system due to marketing agreements between the control system vendor and the I/O manufacturer.

Basic I/O consists of cards with anywhere between 2 to 32 channels dedicated to particular function such as analog inputs (AIs), analog outputs (AOs), thermocouples, RTDs (Resistance Temperature Detectors), discrete inputs (DIs) and discrete outputs (DOs). Wiring can be terminated directly to the card termination block or it can be purchased with remote termination blocks with fusing and a larger footprint to make terminations easier to perform.

Traditional analog I/O transmits the process variable or output via a 4-20 mA 24 VDC signal, where the current is proportional to process variable or output. For example, if the process variable is at 50% of a transmitter's span, the transmitter will output 12 mA. The zero value is elevated so that if a wire is cut or the transmitter loses external power, it can be detected by the control system. The 4-20 mA signal is sent through a 250 Ohm resistor embedded in the I/O card which converts it to a 1-5 VDC signal using Ohm's law. The 1-5 VDC signal is then converted to a digital input by an A to D converter for use by the processer.

Initially, analog I/O only transmitted the signal value, but as control systems evolved, the ability for the controller to react to signal quality and transmitter health drove the need for "smart" I/O. The first "smart" I/O used HART (Highway Addressable Remote Transducer) protocol to propagate transmitter health data. The additional information was transmitted by overlaying a small ripple on the 4-20 mA analog signal and varying the frequency of the ripple to create a digital signal. This allowed the process industry to use signal quality information in their control schemes without upgrading the plant wiring infrastructure.

Many analog transmitters are loop powered, which means the 4-20 mA signal is used to power the transmitter. This implies that the transmitter must be able to function at or slightly below 4 mA. Some transmitters require more power to operate and need an external power source. These transmitters are commonly known as 4 wire transmitters. Four-wire transmitters supply the loop power and are wired to the analog input channel differently than loop powered transmitters; however, this is not always the case. Some 4 wire transmitters have jumpers which allow them to behave like 2 wire transmitters.

Figure 11-4-1 DeltaV Charm 2 Wire Transmitter Wiring

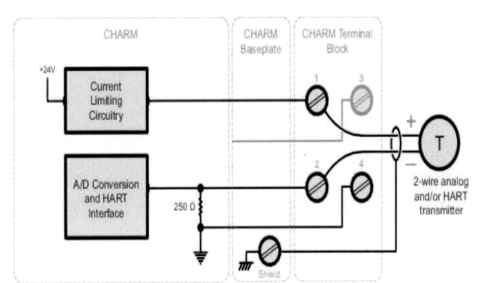

Figure 11-4-2 DeltaV Charm 4 Wire Transmitter Wiring

I/O comes with many different voltage and current ratings. Discrete I/O is usually 120 VAC or 24 VDC. 24VDC I/O is safer for maintenance personnel but is also more susceptible to noise and voltage drop issues. The voltage can come from an external source or be supplied by the control system as is the case with 24 VDC dry contact inputs used with valve limit switches. Analog I/O is typically 24 VDC but some potentiometers require 0 -10 VDC. Solid state outputs are typically rated below 2 Amps and do have some leakage current. For motor circuit outputs, many manufacturers offer relay outputs with current ratings as high as 10 Amps and these provide true isolation. Some manufacturers support thermocouple and RTD temperature inputs eliminating the need for temperature transmitters. If using thermocouple inputs, remember that all of the wiring from the element to the input channel must be made of the same dissimilar metals as the thermocouple.

To reduce wiring costs, sometimes there is a desire to put remote I/O on the plant floor. All of the I/O is terminated locally, and the signals are transmitted digitally over a single cable back to the processor. If this strategy is deployed, make sure the I/O is electrically rated for the area where it is going to be installed. For example, some I/O carries a Class I Division II rating while I/O is only rated for general purpose use. If the I/O does not carry the correct rating, an alternative would be to deploy it in purged cabinets or explosion proof enclosures.

Many manufacturers offer intrinsically safe I/O. Intrinsically safe systems do not contain enough energy to produce ignition if there is a short. Intrinsic safety is achieved by installing isolators or IS barriers in a safe area and installing intrinsically safe transmitters and wiring in the field. The isolators and IS barriers limit the energy available at the device. Some manufacturers incorporate isolators into their I/O cards, eliminating the need for external IS barriers and making the wiring much less complex.

The process control industry has joined the wireless revolution as well. Battery powered wireless transmitters are increasingly being used for non-control applications. The signal is transmitted to a receiver at 2.4 GHz and is secured via a complex encryption join key. Each transmitter acts as a repeater, which means that the more transmitters that are deployed, the more robust the network will become. The receiver can transmit the data to the control system by joining the control system data highway or via OPC (Open Platform Communications) if not supported.

Digital I/O is also available. Digital I/O eliminates error due to voltage drop and due to its speed, allows even more diagnostic information to be transmitted to the control system for maintenance and control. The digital signal is transmitted via a signal cable with multiple devices each having their own address. There are limits on the number of devices that can be on one segment due to distance and power requirements, but in general, bus systems reduce wiring costs. There are many types of digital transmitters and actuators on the market with many different protocols to choose from. Not all control system manufacturers support all of the protocols available. Below is a list of common bus protocols used on field instrumentation.

- ASi-bus

- Profibus

- Fieldbus

- DeviceNet

- Ethernet/IP

Sometimes inputs and outputs to a control system need to come from other foreign control systems, each with their own unique protocols. In these instances, several methods are used. Simple serial communication via RS-232 and RS-485 has been employed for many years. These use open protocols such as Modbus or proprietary protocols from Allen Bradley to transfer data. Today, Ethernet protocol is used much more often to transfer data between systems due to its increased speed.

The PLP uses DeltaV Charm I/O. The advantage of Charm I/O is the elimination of marshalling panels needed for card based I/O. Prior to Charm I/O, inputs and outputs were brought to a marshalling panel which grouped the I/O into individual cards based on voltage and function. Then, multi pair cables were ran from the marshalling panel to the card. Since Charms with different functions can be mixed together, there is no longer the need for marshalling. Electronic marshalling Charms I/O reduces cost and simplifies wiring. From a maintenance standpoint, if an individual channel would fail on a card, the entire card would need to be replaced at a significantly higher cost than replacing an individual Charm.

11.5 Workshop – I/O Identification

Step 11.5.1

From the main graphic Toolbar, open DeltaV Explorer. Expand the System Configuration and drill down to module FIC-106. Click on FIC-106, then right click on the AI1$IO_IN parameter and select properties.

Graphic 11-5-1 DeltaV Control Studio Launch

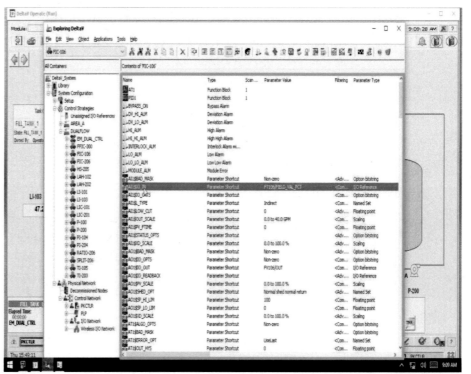

Step 11.5.2

Once the properties of the AI1$IO_IN parameter appear, record the Device Tag and Device Tag Parameter that the module is using. Record all of the available Device Tag Parameters.

Device Tag = _____

Device Tag Parameter = _____

_____

_____

_____

_____

Graphic 11-5-2 DeltaV Device Tag

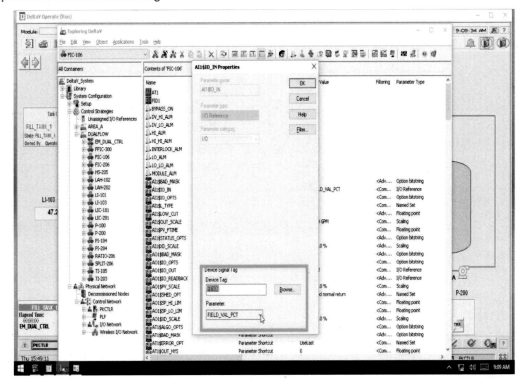

Step 11.5.3

Launch Control Studio On-Line for FIC-106. The Charm address the input is assigned to will appear under the AI1 block. Record this address.

FIC-106 Input Charm Address = _____

Graphic 11-5-3 DeltaV Control Studio On-Line Access

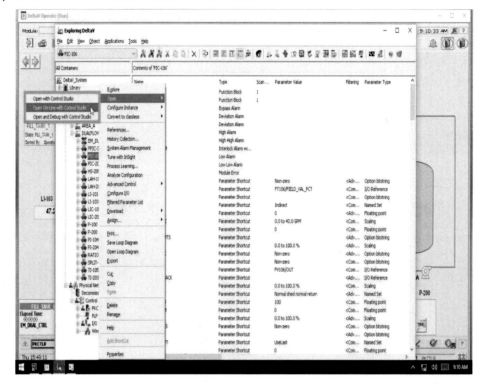

Step 11.5.4

Return to DeltaV Explorer, expand the Physical Network and drill down to the Charm recorded in step 11.5.3. Right click on the Charm, select properties and record the Charm type.

FIC-106 Input Charm Type = _____

Graphic 11-5-4 FIC-106 Charm Type

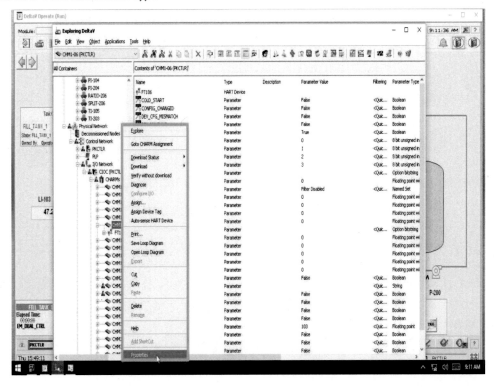

11.6 Workshop – I/O Scaling

Step 11.6.1

From the main graphic, Startup the PLP using a complexity setting of Flow Control. *You may need to fill tank TK-100 using the FILL TANK 1 sequence.*

Step 11.6.2

Using what you learned from the previous workshop, determine the input address of FIC-206 and record it. From the main graphic, launch DeltaV Diagnostic and expand the I/O network. In the right pane, double click on the associated FIC-206 input Charm, scroll down and record the Charm input Value, the PV on DeltaV and the transmitter reading on the PLP.

FIC-206 Input Charm Address = _____

FIC-206 Input Charm Value = _____

FIC-206 PV on DeltaV = _____

FIT-206 Transmitter reading = _____

Graphic 11-6-1 DeltaV Diagnostic Charm Value Access

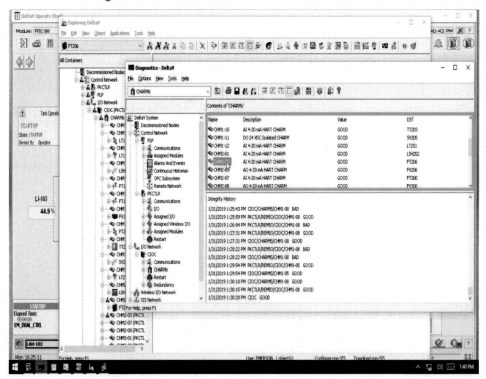

Step 11.6.3

Return to DeltaV Explorer. Drill down to the associated FIC-206 input Charm, right click on it and select Configure to launch AMS. In AMS, go to Manual Setup and on the Basic Setup tab, change the PV URV from 40 GPM to 50 GPM. It will highlight in yellow. Click on the Send button to send to the transmitter. Allow FIC-206 flow to stabilize.

Graphic 11-6-3 AMS Upper Range Value Change

Step 11.6.4

Return to DeltaV Diagnostics and using the instructions in step 11.6.2, record the input field Value of FIC-206, the PV on DeltaV and the transmitter reading. Are they different and why? Which reading is correct?

FIC-206 Input Charm Value = _____

FIC-206 PV on DeltaV = _____

FIT-206 PV on Transmitter = _____

_____

_____

_____

Step 11.6.5

Return the transmitter PV URV to 40 GPM using the procedure in step 11.6.3.

11.7   Workshop – Analog Input Wiring

Step 11.7.1

Using what you learned from the previous workshops, determine the address of the input Charms for FIC-106 and FIC-206. Go to the associated transmitters on the PLP and open up the wiring bay for each transmitter. Which transmitter is a 2 wire transmitter and which one is a 4 wire transmitter?

2 Wire Transmitter = _____

4 Wire Transmitter = _____

Step 11.7.2

Go to the PLP I/O cabinet and remove the associated input Charms for FIC-106 and FIC-206 by pushing the associated orange lever. Record the model number of each Charm. Are the Charms the same?

FIC-106 Input Charm Model Number = _____

FIC-206 Input Charm Model Number = _____

_____

_____

Step 11.7.3

Observe the termination on the inputs for FIC-106 and FIC-206. Are they the same? If so, why are they based on figures 11-4-1 and 11-4-2?

_____

_____

_____

Step 11.7.4

Remove the termination blocks for the inputs to FIC-106 and FIC-206 by pushing on the gray release button on the Charm base plate to the left of each of the termination blocks.   Record the termination block model numbers for the inputs to FIC-106 and FIC-206. Are they the same?

FIC-106 Charm Termination Block Model Number = _____

FIC-206 Charm Termination Block Model Number = _____

_____

_____

_____

Step 11.7.5

Reinstall the termination blocks and Charms, and verify the transmitters are communicating with DeltaV.

11.8   Conclusions

There are many types of I/O available for purchase, each requiring unique wiring. Some require external jumpers on the termination blocks to complete the loop circuit or supply power. Always consult the manufacturer's wiring diagrams during the design phase to avoid commissioning delays. If using traditional analog I/O, the span of the transmitter and the control system must match to allow the control system to interpret the 4-20 mA signal correctly. Not all control systems integrate easily with all types of I/O, which is why I/O requirements should be considered in the control system selection process.

For more information on Emerson training classes visit the following website:

https://www.emerson.com/en-us/automation/services-consulting/educational-services

Section 12:   Control Networks

12.1   Objectives

When the student has completed this module, the student will:

- Be able to describe how the DeltaV Network is setup.

- Understand the security risks associated with control system networks.

- Understand DeltaV network redundancy.

12.2   Intended Audience

Instrument Technician

Process Engineer

12.3   Prerequisites

Section 2 PLP Infrastructure

Section 3 PLP Instrumentation

Section 4 DeltaV Navigation

Section 5 DeltaV Modes and Signal Status

12.4   Discussion

A control system's network is the way data is transferred for the I/O to the controller and from the controller to the operator interface. Some control system networks are proprietary, but many use Ethernet communication that is no different than a typical office LAN (Local Area Network) with the exception of network redundancy and shielded cable. Control system networks are subject to the same architecture rules with respect to cable lengths, speed and security as the office LAN.

On an Ethernet network, each device is assigned an IP (Internet Protocol) address and packets of data are sent out with a destination address as the need arises. Devices on the network are constantly pulling down packets of information with their IP address on them. If the network does not have the proper bandwidth for all this information, collisions and network problems can occur.

Networks are layered to minimize traffic and to improve security due to firewalls where the layers interact. Layered networks ensure that only certain types of devices see only certain types of data. Figure 12-4-1 is an example of a typical DeltaV network.

Figure 12-4-1 Typical DeltaV Network

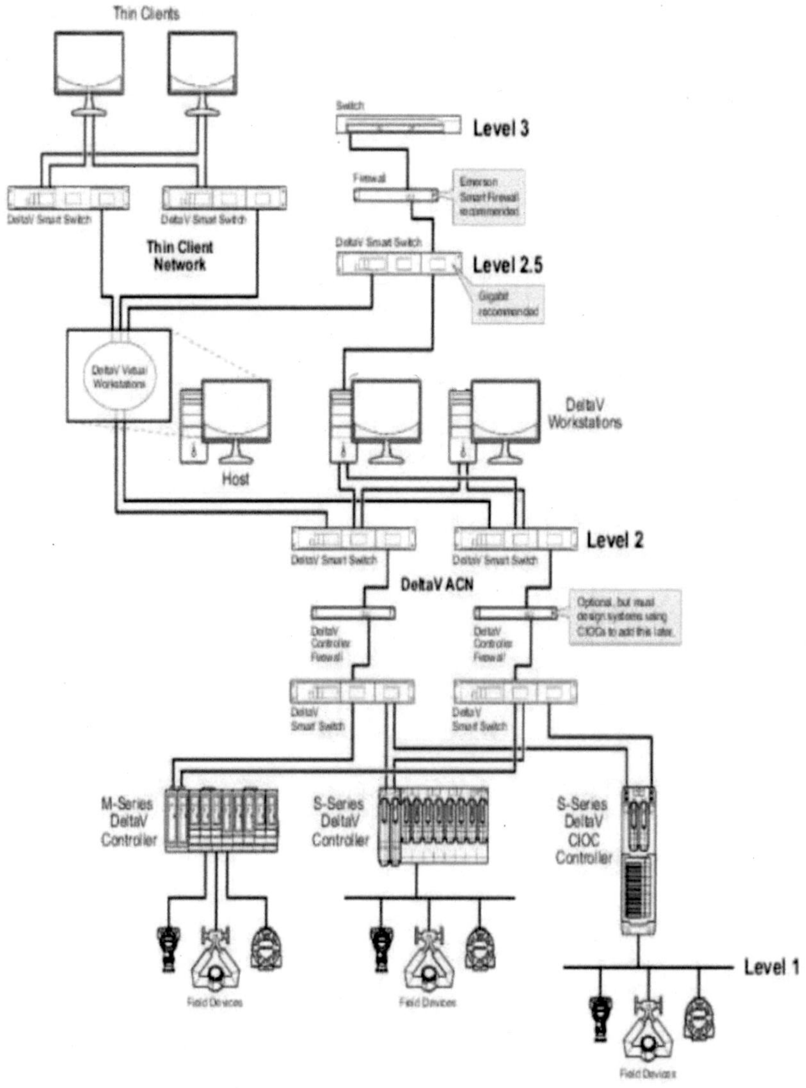

Layer 2 is redundant for network robustness. Different physical paths are recommended for the primary and secondary networks to minimize risks associated with the network's ambient environment.

Data for production reports and the Enterprise continuous data historian is extracted from the control system at Layer 3. A firewall is essential in this location to prevent viruses and malware from propagating down to the control system. Remember that every USB port on every operating workstation is a potential security breech point. Always lock or disable USB ports when not in use.

12.5   Workshop – Network Diagnostics

Step 12.5.1

From the main graphic Toolbar, launch DeltaV Diagnostics. In the Control Network, drill down to the PKCTLR controller, right click and select TCP Ping. Record the controller's primary address and the time it took to do the ping.

PKCTLR Primary IP Address = _____

PKCTLR Primary Ping Time = _____

Graphic 12-5-1 DeltaV Controller Ping

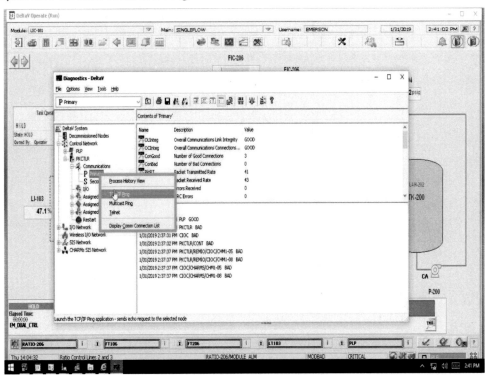

Step 12.5.2

Repeat step 12.5.1 for the secondary connection.

PKCTLR Secondary IP Address = _____

PKCTLR Secondary Ping Time = _____

12.6 Workshop –Network Redundancy

Step 12.6.1

From the graphic screen, set the Complexity to Flow Control and Tank Operation to STARTUP. PLP will startup. Allow FIC-106 and FIC-206 to stabilize at around 5.0 GPM. *You may need to run Fill Tank 1 first to get TK-100 above 55%.*

Graphic 12-6-1 PLP Startup

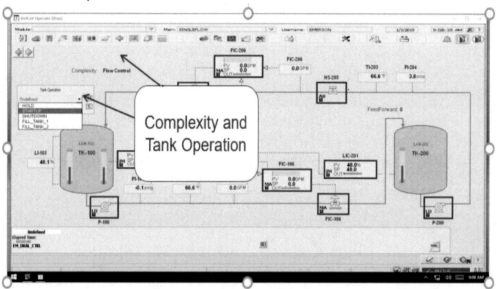

Step 12.6.2

Go to the PLP I/O cabinet and locate the Primary Ethernet (SW1) with the yellow cables landing on it. Pull the yellow Ethernet cable going to the operator workstation. Record any alarms that appear on the main graphic in the Alarm Banner. Does the PLP continue to run?

Alarms = _____

_____

_____

Step 12.6.3

Go to the PLP I/O cabinet and locate the Secondary Ethernet (SW2) with the black cables landing on it. Pull the black Ethernet cable going to the operator workstation. What happens to the main graphic? Does the PLP continue to run?

_____

_____

_____

Step 12.6.4

Plug the workstation cables back in. Restart the PLP, if needed, and allow the flows to stabilize.

Step 12.6.5

Go to the PLP I/O cabinet and locate the Primary Ethernet (SW1) with the yellow cables landing on it. Pull the yellow Ethernet cable going to the PK controller. Record any alarms that appear on the main graphic in the Alarm Banner. Does the PLP continue to run?

Alarms = _____

_____

_____

Step 12.6.6

Go to the PLP I/O cabinet and locate the Secondary Ethernet (SW2) with the black cables landing on it. Pull the black Ethernet cable going to the PK controller. What happens to the main graphic? Does the PLP continue to run?

_____

_____

_____

Step 12.6.7

Plug the workstation cables back in. Restart the PLP, if needed, and allow the flows to stabilize. *Note: You may have to cycle power to the PLP to recover from this test.* What happened when you plugged both cables back in?

_____

_____

_____

Step 12.6.8

Go to the PLP I/O cabinet and locate the Primary Ethernet (SW1) with the yellow cables landing on it. Pull the yellow Ethernet cable going to the CIOC. Record any alarms that appear on the main graphic in the Alarm Banner. Does the PLP continue to run?

Alarms = _____

_____

_____

Step 12.6.9

Go to the PLP I/O cabinet and locate the Secondary Ethernet (SW2) with the black cables landing on it. Pull the black Ethernet cable going to the CIOC. What happens to the main graphic? Does the PLP continue to run?

_____

_____

_____

Step 12.6.10

Plug the workstation cables back in. Restart the PLP, if needed, and allow the flows to stabilize. *Note: You may have to cycle power to the PLP to recover from this test.* What happened when you plugged both cables back in?

_____

_____

_____

12.7   Conclusions

A control system's network must be robust to prevent unexpected production interruptions. One way to accomplish this is through redundancy. It is important that the physical network is laid out in such a way as to minimize common faults due to weather and/or construction mishaps.

Since DeltaV uses the same network technology as the plant LAN, a control system security protocol should be developed and periodically reviewed as additional threats are encountered or equipment is added.

For more information on Emerson training classes visit the following website:

https://www.emerson.com/en-us/automation/services-consulting/educational-services

Section 13:   Troubleshooting

13.1   Objectives

When the student has completed this module, the student will:

- Understand some of the common sources of device failures.

- Be able to identify common failures using DeltaV diagnostic tool.

13.2   Intended Audience

Instrument Technician

Process Engineer

Operator

13.3   Prerequisites

Section 2 PLP Infrastructure

Section 3 PLP Instrumentation

Section 4 DeltaV Navigation

Section 5 DeltaV Modes and Signal Status

13.4   Discussion

Production facilities consist of a vast amount of complex equipment which is constantly in service. Even with preventative maintenance programs, equipment still fails due to fatigue. The plant control system usually provides the first indication that a failure has occurred, and it is the primary tool used to identify the time and cause of the failure. Some root causes of downtime due to equipment failures are:

- Corrosion – This can be the result of moisture in the air, standing water or small leaks in the piping system.

- Loose Wires – This results from thermal expansion and contraction over time.

- Vibration – Large rotating equipment is constantly causing movement that will work terminations or other fasteners loose. These are some of the most difficult faults to find since they are intermittent.

- Electronic – Solid state components can fail due to thermal cycling. This is accelerated in a plant environment.

A large percentage of production problems have nothing to do with production equipment failing at all. The root cause may be the result of an external issue. Examples include:

- Loss of Air – The plant instrument air system goes down due to a compressor failure or a manual valve is left closed by mistake after servicing equipment.

- Loss of Electricity – The electric utility has a weather-related interruption or a fault due to the age of the infrastructure.

- New Construction – While adding or demolishing wiring inside a control system cabinet, an electrician accidently bumps a wire or cuts the wrong wire.

- Operator Error – An operator may accidently push the wrong button on an operating console. This could easily happen in a stressful situation such as the startup of a continuous production facility.

Failures typically do not happen when someone is there to witness it. Process engineers and maintenance personnel need to rely heavily on the control system's historical tools to recreate the failure scenario. On the PLP, the two DeltaV tools needed to do this are the DeltaV Continuous Historian and the DeltaV Event Journal. Both of these tools can be accessed with the Process History View application on the Toolbar.

- DeltaV Continuous Historian – This tool provides the ability to create a historical trend of a parameter if it has been configured for historical logging. Out of the box, DeltaV gives the user the ability to collect data on 250 parameters and, if needed, the user can increase this number with additional licensing.

- DeltaV Event Chronicle – This tool logs all of the alarms that happen and all of the operator entries that have been made. The tool has the ability to filter events based on time, area and module name, making it easier to sift through the vast amount of data collected.

Although not part of the PLP, DeltaV also has a Batch Historian tool available. When DeltaV batch is deployed, the Batch Historian will capture the recipe values used, operator responses and the start and end times of phases, unit procedures and procedures.

13.5 Workshop – Loose Wire

Step 13.5.1

From the graphic screen, set the Complexity to Flow Control and Tank Operation to STARTUP. PLP will startup. Allow FIC-106 and FIC-206 to stabilize at around 5.0 GPM. *You may need to run Fill Tank 1 first to get TK-100 above 55%.*

Graphic 13-5-1 PLP Startup

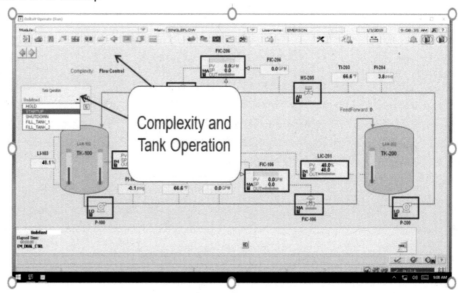

Step 13.5.2

Go to the PLP I/O cabinet and locate CHM2-02 for input FIT-206. Remove the wire on terminal 1 of CHM2-02. Does the PLP continue to run? What alarms appear in the Alarm Banner? What is the process variable reading on FIC-206?

_____

_____

_____

Step 13.5.3

Click on FIC-206 on the main graphic to bring up the faceplate. What happened to the background color of the process variable? Click on the I/O Diagnostic Detail. What message is it displaying?

_____

_____

_____

Step 13.5.4

Click on the H in the Tank Operation box on the main graphic. This provides the operator with
information on what caused the sequence to shut down. What was the module identified as the
First Out Failure?

_____

_____

_____

Graphic 13-5-4 Equipment Module Hold Monitor Access

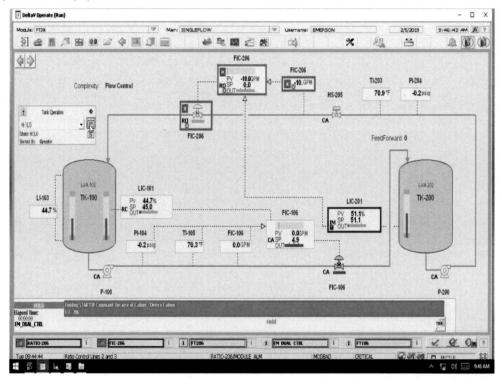

Step 13.5.5

From the main graphic Toolbar, launch the Process History View application. Create a custom
Event trend called FIC-206 Failure and save it. Clear the filters under the Event menu selection.

Graphic 13-5-5 DeltaV PHV Event Filtering

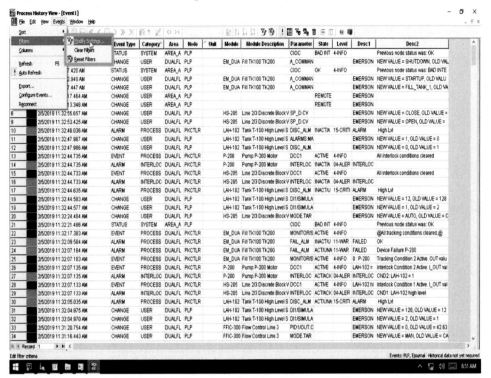

Step 13.5.6

Modify filter settings by narrowing the search under the Module Tab to only look for FIC-206 and under the Date/Time tab, set the time span close to the failure time. Record the time the wire was removed.

_____

_____

_____

Graphic 13-5-6 Modifying Filter Settings

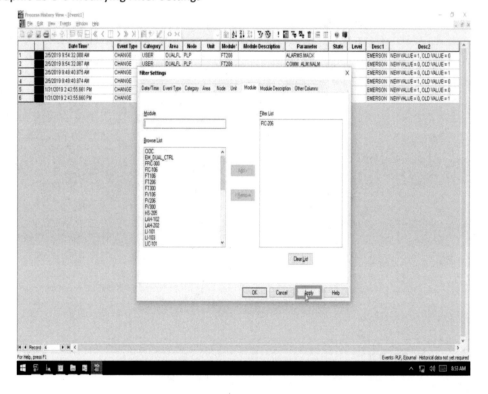

Step 13.5.7

Go back to the main graphic and click on FIC-206 to bring up the faceplate. At the bottom of the faceplate, click on the PHV trend. Go back to the time on the trend to when the PV registered low. In the events below the trend, find the exact time of the Bad PV alarm. Does it match up with the timestamp recorded in step 13.5. 5?

_____

_____

_____

Step 13.5.8

Reconnect the wire on CHM2-02.

13.6   Workshop – Loss of Instrument Air

Step 13.6.1

From the graphic screen, set the Complexity to Flow Control and Tank Operation to STARTUP. PLP will startup. Allow FIC-106 and FIC-206 to stabilize at around 5.0 GPM. *You may need to run Fill Tank 1 first to get TK-100 above 55%.*

Step 13.6.2

On the PLP, turn the manual valve supplying air to the unit off. Does the PLP continue to run? Does the output of FIC-206 match the true valve position on FV-206 on the PLP?

_____

_____

_____

Step 13.6.3

Click on the H in the Tank Operation box on the main graphic. This provides the operator with information on what caused the sequence to shut down. What was the module identified as the First Out Failure?

_____

_____

_____

Step 13.6.4

Click on the faceplate of the module identified as the First Out Failure in step 13.6.3. Click on the Interlock Detail on the faceplate. Record the condition with the red arrow that caused the module to fail. Is the module still interlocked? Why is it no longer interlocked?

_____

_____

_____

Step 13.6.5

Open the manual air valve feeding the PLP.

13.7 Workshop – Loss of Power

Step 13.7.1

From the graphic screen, set the Complexity to Flow Control and Tank Operation to STARTUP. PLP will startup. Allow FIC-106 and FIC-206 to stabilize at around 5.0 GPM. *You may need to run Fill Tank 1 first to get TK-100 above 55%.*

Step 13.7.2

Go to the PLP I/O cabinet and pull fuse X6-F04 that feeds power to CIOC baseplate 1. Does the PLP keep running? What alarms appear on the Alarm Banner?

_____

_____

_____

Step 13.7.3

Click on the H in the Tank Operation box on the main graphic. This provides the operator with information on what caused the sequence to shut down. What was the module identified as the First Out Failure?

_____

_____

_____

Step 13.7.4

Click on the faceplate of the module identified as the First Out Failure in step 13.7.3. Click on the Interlock Detail on the faceplate. Record the condition with the red arrow that caused the module to fail. Is the module still interlocked?

_____

_____

_____

Step 13.7.5

Reinstall fuse X6-F04 to CIOC baseplate 1. Acknowledge all of the alarms and restart the PLP allowing the flows to stabilize.

Step 13.7.6

Go to the PLP I/O cabinet and pull fuse X6-F12 that feeds power to CIOC baseplate 2. Does the PLP keep running and if so, why? What alarms appear on the Alarm Banner?

_____

_____

_____

Step 13.7.7

Reinstall fuse X6-F04 to CIOC baseplate 1. Acknowledge all of the alarms and restart the PLP if necessary. Allow the flows to stabilize.

Step 13.7.8

Go to the PLP I/O cabinet and pull fuse X6-F01 that feeds power to the CIOC card holder baseplate. Does the PLP keep running? What alarms appear on the Alarm Banner?

_____

_____

_____

Step 13.7.9

Go to the PLP I/O cabinet and pull fuse X6-F09, the redundant power to the CIOC card holder baseplate. Does the PLP keep running? What alarms appear on the Alarm Banner?

_____

_____

_____

Step 13.7.10

Click on the H in the Tank Operation box on the main graphic. This provides the operator with information on what caused the sequence to shut down. What was the module identified as the First Out Failure?

_____

_____

_____

Step 13.7.11

On the main graphic, click on FIC-206 to bring up the faceplate. Does the output match the true valve position of FV-206 on the PLP? Does the PV match the transmitter FIT-206 on the PLP? Why do they not match?

_____

_____

Step 13.7.12

From the FIC-206 faceplate, launch Control Studio On-Line and in the diagram, browse to the AI1 function block. What is displayed on the AI1/OUT parameter?

_____

_____

_____

Step 13.7.13

Reinstall fuses X6-F01 and X6-F09.

13.8   Conclusions

Troubleshooting skills are needed by both production and maintenance personnel. Operators and Process Engineers need to be able to describe the source of the issue to maintenance, so they have an idea where to look. Sometimes failures work in a cascade scheme and the troubleshooter needs to follow the path downward to get to the root cause of the issue.

A good understanding of the capabilities of trend tools and alarm logs are critical to tracking down the root cause of any failure. Most of the time operators or engineers are not present when a failure happens so historical applications are the only way to reconstruct the sequence of events related to the failure.

For more information on Emerson training classes visit the following website:

https://www.emerson.com/en-us/automation/services-consulting/educational-services

Section 14: Grounding and Shielding

14.1 Objectives

When the student has completed this module, the student will:

- Understand the importance of grounding for safe operations.
- Understand the importance of shielding.
- Understand and be able to describe the effects of ground loops in instrument wiring.
- Understand and be able to deploy good grounding practices.

14.2 Intended Audience

Instrument Technicians

14.3 Prerequisites

Section 2 PLP Infrastructure

Section 3 PLP Instrumentation

Section 4 DeltaV Navigation

Section 5 DeltaV Modes and Signal Status

14.4 Discussion

14.4.1 Grounding

Proper grounding is essential for a control system to be safe and work properly. There are several things to consider with respect to grounding.

- Safety Grounding (AC Power Grounding) – This protects personnel from shock hazards when insulation fails or current carrying conductors are shorted to ground. The ground conductor must be sized to handle the entire fault current. When there is a short, fault current will follow the path of least resistance and this must be the ground system and not through other equipment or personnel. If there is minimal resistance due to a good ground path, the resulting overcurrent situation will blow a fuse or open a circuit breaker and clear the fault.

- High Frequency Grounding – The ground system should be able to dissipate noise from equipment such as variable frequency drives and welders. Items such as these create electromagnetic interference and induce voltages which can lead to noise on low voltage instrument wiring.

- DC Signal or DC Reference Grounding – For a control system to work properly, all of the low voltage signal wiring needs to have a common potential. Elevated or different zero potentials can lead to faulty readings. A good DC ground should be less than 3 Ohms when measured from the triad ground or grid ground to earth. This ground should be kept separate from the AC power ground until it is close to earth potential. The purpose of this is to prevent the DC system zero potential from elevating when there is a short on the AC system. If the grounds do not combine until close to earth potential, utility faults and lighting strikes will have little effect on the low voltage control wiring.

- Lighting Safety Grounding – When lighting strikes a facility, it creates a large voltage gradient and the resulting current will seek earth ground to dissipate this voltage. Lighting will strike the tallest point in a facility. If the tallest point is tied to building steel and the building steel is properly grounded, the current will go directly to the earth and not though personnel or other equipment.

Figure 14-4-1 Typical Charm Cabinet Grounding

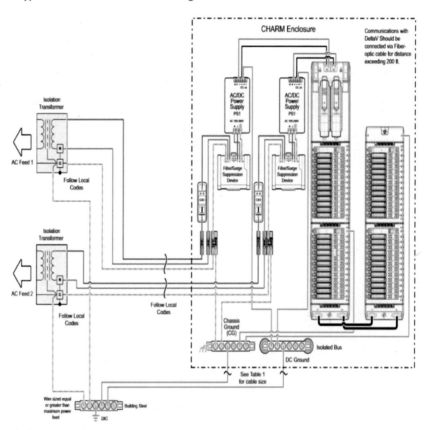

14.4.2   Shielding

Shielding is important to have minimal noise on DC signal wiring. Electromagnetic Interference (EMI) is everywhere and there are many loads in a production facility that can induce a voltage on neighboring wiring. By shielding and grounding a signal cable, the EMI problem can be mitigated.

It is important to only ground the cable on one side to prevent ground loops from forming. If the zero potential is different at one end of a cable as compared to the other, the difference in potential will cause current to flow and noise. Another recommended practice is to ground the cable shield at the end from where the loop power is being sourced. This makes sense since the loop zero potential point is at the device on a 4 wire transmitter.

14.5 Workshop – Signal Cable Shielding

Step 14.5.1

From the graphic screen, set the Complexity to Flow Control and Tank Operation to STARTUP. PLP will startup. Allow FIC-106 and FIC-206 to stabilize at around 5.0 GPM. *You may need to run Fill Tank 1 first to get TK-100 above 55%.*

Graphic 13-5-1 PLP Startup

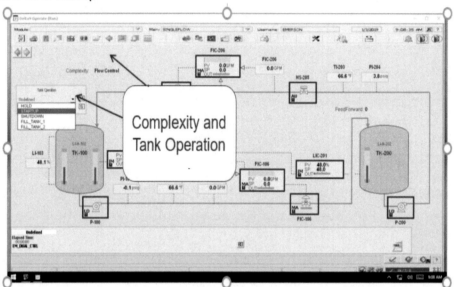

Step 14.5.2

From the main graphic, click on PI-204 to bring up the faceplate. Launch Process History View from the Icon at the bottom of the faceplate. Rescale the Y axis of the trend by right clicking on the Parameter Reference on the bottom of the trend. Set the Y axis scale so it is 0.5 PSIG above and below the normal operating pressure.

Graphic 14-5-2 Trend Y Axis Rescale

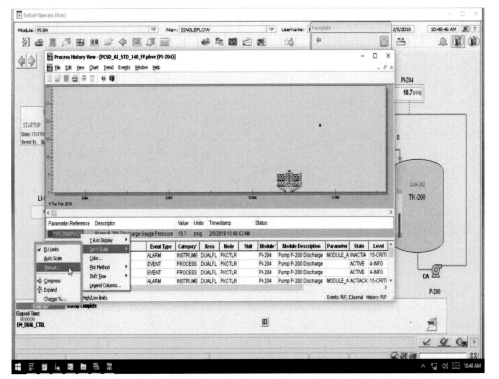

Step 14.5.3

Using the PHV Toolbar, change the time scale of the trend to about 15 minutes.

Step 14.5.4

Go to the PLP I/O cabinet. Locate Charm CHM1-08 for PIT-204 and remove the shield for that cable from the Charm holder baseplate.

Step 14.5.5

Take a large AC load and place it next to the PIT-204 cable at the transmitter and run the load for 15 minutes observing the PIT-204 trend.

Step 14.5.6

Turn off the AC load and re-terminate the PIT-204 cable shield. Observe the trend on PIT-204. Did you see any difference?

_____

_____

_____

14.6   Workshop – AC and DC Grounding

Step 14.6.1

Remove power from the PLP by turning off the two circuits feeding the unit.

Step 14.6.2

With the power off, remove the two green AC ground wires and from the PLP feed circuits on the bottom terminal strip. Once the grounds are removed, re-energize the two PLP feed circuits. Allow the PLP to boot up and logon to the system.

Step 14.6.3

From the graphic screen, set the Complexity to Flow Control and Tank Operation to STARTUP. PLP will startup. Allow FIC-106 and FIC-206 to stabilize at around 5.0 GPM. *You may need to run Fill Tank 1 first to get TK-100 above 55%.*

Step 14.6.4

From the main graphic, click on PI-204 to bring up the faceplate. Launch Process History View from the Icon at the bottom of the faceplate. Rescale the Y axis of the trend by right clicking on the Parameter Reference on the bottom of the trend. Set the Y axis scale so it is a 0.5 PSIG above and below the normal operating pressure.

Step 14.6.5

Using the PHV Toolbar, change the time scale of the trend to about 15 minutes.

Step 14.6.6

Allow the PLP to run for 5 minutes. At 5 minutes, do a screen capture of the PI-204 trend.

Step 14.6.7

Shut down the PLP with the Tank Operation equipment module.

Step 14.6.8

Turn off the two circuits to the PLP and re-terminate the two green ground wires.

Step 14.6.9

Repeat steps 14.6.3 through 14.6.6. Did you notice any difference in the two trends of PIT-204?

_____

_____

_____

Step 14.6.10

Remove the jumper between the AC and DC ground bars in the middle of the PLP I/O cabinet and repeat step 14.6.3 through 14.6.6. Did you notice any difference in the PIT-204 trend?

_____

_____

_____

Step 14.6.11

Re-terminate the jumper between the AC and DC ground bars.

14.7   Conclusions

AC power grounding is essential for personnel and equipment protection. Grounding should be inspected from time to time to make sure it is intact. Failure to do so could result in injury or death.

DC grounding creates the zero potential for all measurements. This should be separated from the AC power ground until it is near earth potential. Elevated or multiple DC grounds will result in noise on the system.

All instrument signal cable should be shielded to prevent noise on the signal due to EMI. The shield should be grounded at one of the cable, and on the other end, it should be taped back and insulated to prevent ground loops.

For more information on Emerson training classes visit the following website:

https://www.emerson.com/en-us/automation/services-consulting/educational-services

Section 15:   Interlocks

15.1   Objectives

When the student has completed this module, the student will:

- Be able to describe the function and importance of interlocks.

- Understand how discrete control devices respond when interlocked.

15.2   Intended Audience

Instrument Technicians

Process Engineers

Operators

15.3   Prerequisites

Section 2 PLP Infrastructure

Section 3 PLP Instrumentation

Section 4 DeltaV Navigation

Section 5 DeltaV Modes and Signal Status

15.4   Discussion

During the design stage of any facility, a HAZOP (Hazard and Operability) study should be conducted to identify all of the potential hazards in the process. This study should be reviewed on a periodic basis and updated if any major changes are done to the process. Once the hazards have been identified, a LOPA (Layers of Protection Analysis) should be developed to identify all of the independent Layers of Protection needed to minimize any hazard. An interlock could be one of the Layers of Protection required. For example, an interlock could be to close the feed valve if the level in a tank gets too high.

Interlocks consist of three parts: the sensor, the logic solver and the final control element. The number of sensors, logic solvers and final control elements and their configurations are dependent on the quantitative risk factor developed for a hazard and the acceptable risk the operating entity is willing to accept. Depending on how the interlock fits into the SIS (Safety Instrumented System) plan, it may be part of the BPCS (Basic Process Control System) or it may be implemented with a dedicated and independent SIS logic solver. The DeltaV system on the PLP is an example of a BPCS. Typically, the BPCS will contain interlocks that mimic the actions of the SIS system as an added layer of protection.

Interlocks should be monitoring the process 100% of the time and configured in a way so that batch sequencing or operators should not have to bypass the interlock during normal operations. In the rare occasion that an interlock needs to be bypassed for maintenance reasons, a change control should be written and the bypass logged. The interlock bypass log book should be reviewed at the start of each shift.

Interlocks should be tested on a periodic basis to make sure they are functioning correctly. Some final control elements are seldom actuated and could become frozen in the open position. Instruments that fall out of calibration or code that is changed can inadvertently change the function of an interlock. It is recommended that interlocks be tested as a loop versus individual components, but this is not always possible.

In DeltaV, when a DCD (Discrete Control Device) is interlocked, the DCD output is locked in the passive State and the actual mode will be locked in LO (Local Override), preventing any sequencer or operator from changing States. If the device is in the active State at the time the trigger condition comes true, a device fail alarm will be generated.

When an interlock is triggered on a DCD, the device will typically stay in the passive State even after the interlock clears; however, this is configurable. Anytime an interlock is configured, the control system should be setup so that it captures the cause of the device failure. In DeltaV, this is done on the DCD Interlock Detail with a red arrow. Sometimes, multiple interlocks will be triggered during a process interruption. In these instances, it is important to do a first out trapping on the trigger conditions so the root cause can be determined.

PID loops can also be interlocked as well. When a PID loop is interlocked, the actual mode will be locked in LO preventing batch sequencers or operators from changing the loop output, and the loop output will be sent to the TRK_IN value. There is the ability, via configuration, to give the operator override functionality in MAN mode if desired. Typically, once the trigger condition clears, the PID action will take over and drive the loop to setpoint.

15.5   Workshop – Discrete Control Device Interlock

Step 15.5.1

From the graphic screen, set the Complexity to Flow Control and Tank Operation to STARTUP. PLP will startup. Allow FIC-106 and FIC-206 to stabilize at around 5.0 GPM. *You may need to run Fill Tank 1 first to get TK-100 above 55%.*

Graphic 15-5-1 PLP Startup

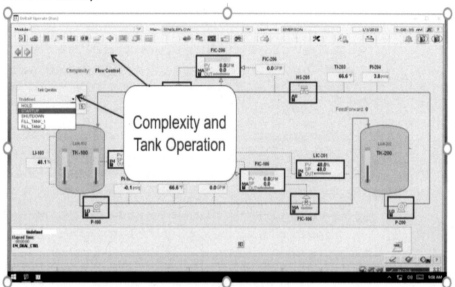

Step 15.5.2

Click on LAH-102 to bring up the faceplate. Click on the I/O Detail Icon to bring up the I/O Detail. Simulate a high level on TK-100 by clicking the Simulate check box. Did the PLP keep running? What is the actual mode of HS-205? Can you change the actual mode? Can you change the requested setpoint? What was the first out condition on HS-205 on the Interlock Detail? Did HS-205 fail and if so, why?

_____

_____

_____

Graphic 15-5-2 LAH-102 High Level Simulation

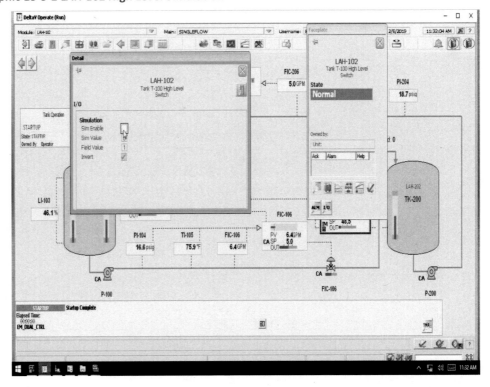

Step 15.5.3

Try and Startup the PLP again from the Tank Operation box on the main graphic. Will the PLP run?

_____

_____

_____

Step 15.5.4

Remove the Simulation on LAH-102. Go to the faceplate on HS-205. What is the State of the valve? Can you change the mode to AUTO and open it?

_____

_____

_____

15.6   Conclusions

Interlocks are critical for the safe operation of a facility and should be tested on a periodic basis. The equipment used to implement an interlock is dependent on the amount of risk associated with the hazard it is trying to mitigate. When configuring an interlock, thought should be given to the recovery process once the trigger condition is cleared to prevent additional process upsets.

For more information on Emerson training classes visit the following website:

https://www.emerson.com/en-us/automation/services-consulting/educational-services

Section 16:   PID Loop Tuning

16.1   Objectives

When the student has completed this module, the student will:

- Be able to describe the difference between direct and reverse acting loops.

- Be able to describe when to use P, PI and PID control.

- Be able to describe the difference between self-regulating, integrating and runaway processes.

- Understand basic loop tuning concepts and terms.

- Be able to tune loops using Ziegler-Nichols open loop tuning and Lambda tuning methods.

16.2   Intended Audience

Instrument Technicians

Process Engineers

Operators

16.3   Prerequisites

Section 2 PLP Infrastructure

Section 3 PLP Instrumentation

Section 4 DeltaV Navigation

Section 5 DeltaV Modes and Signal Status

16.4   Discussion

The Proportional-Integral-Derivative (PID) controller is the primary tool used in the process industry to control flow, temperature, level, pressure and composition. Tuned properly, the PID controller can accomplish almost all of the basic control functions in the plant. A function block representation of a feedback loop is shown in Figure 16-4-1. The controller $g_c(s)$ is adjusting its output based on the feedback from the measured variable and the controller algorithm.

Figure 16-4-1 Feedback Loop Block Diagram Representation

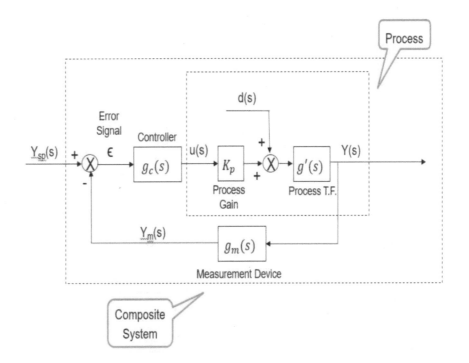

A controller based on the ISA Standard Form with action on error consists of three components: Proportional, Integral and Derivative, each being affected by the tuning factors K_c (gain), T_i (reset) and T_d (rate). The equations are shown in Figure 16-4-2. Some important observations can be made by examining these equations.

- The fact that the equations are based on the ISA Standard Form implies that not all PIDs algorithms are the same. Different manufacturers use different equations. In fact, in DeltaV this is user configurable.

- Units are important since not all manufacturers use the same algorithms. Some use seconds while others use minutes for rate and reset. Some use repeats per minute or repeats per second for reset. The gain factor is unitless and should be calculated using percent of span of the process variable. Some controllers use proportional band which is equal to $100/K_c$.

- Controller gain K$_c$ affects all three components.

- The Integral component is based on the integral of the error (area under the curve between setpoint and the process variable). The larger the error, the bigger impact the integral component will have. As long as there is some error, the integral component will continue driving the process to setpoint. This can lead to overshoot, and depending on the process, the results can be undesirable if the reset parameter is too aggressive. Many times on reactor temperature loops, the reset parameter is de-tuned to prevent any overshoot.

- The Derivative component is based on the rate of change of the error (velocity at which the process variable is moving to setpoint). If the process variable measurement is noisy and the rate component is set too high, it will amplify the noise and lead to instability. Therefore, the rate parameter is set to zero in flow applications.

Figure 16-4-2 ISA Standard Form PID Equations

Proportional P = $K_c(E)$

Integral I = $K_c\left(\frac{1}{Ti}\right)\int E dt$

Derivative D = $K_c(T_d)\frac{dE}{dt}$

Where

E = error (PV − SP) %

SP = Setpoint in %

PV = Process Variable in %

P = Proportional mode contribution %

I = Integral mode contribution in %

D = Derivative mode contribution in %

K_c = Controller gain (no units)

T_i = Integral time (reset)in seconds

T_d = Derivative time (rate) in seconds

When configuring a PID loop, the loop action needs to be determined. Direct acting controllers changes the controller output in the same direction as the error signal. When the error increases, the controller output increases. For example, if a control valve is being used to control level on the discharge piping of a tank, this controller action would be setup as direct. In other words, as the tank level moves above setpoint creating a positive error, the valve would need to open (increase) the controller output to bring the level back to setpoint.

Reverse acting controllers changes the controller output in the opposite direction as the error signal. For example, if a control valve is in series with the flow transmitter and the flow increased creating positive error, the controller output would need to decrease to bring the flow back to setpoint. Controller action is independent of fail action of the valve. If the control valve is fail open, simply invert the PID output prior to sending it to the field.

Before tuning a loop, it is important to know the objective of the tuning and the open loop process response. In some tuning methods, different equations are used to calculate gain, reset and rate depending on whether it is a P, PI or PID controller. The control engineer may choose less aggressive or more aggressive tuning settings than the tuning method produces. Some common tuning objectives are:

- Minimum PV overshoot of SP – Critical for some reactor temperature reactions and compressor surge.

- Maximum absorption of variability – Using a tank level as a surge tank to absorb process upsets by not passing the disturbance on to outlet flows.

- Minimum PV integrated error in load or setpoint response – Reduce the amount of off-spec product being produced.

- Minimum PV peak error in a load response – Prevent compressor surge, pressure relief or SIS activation.

There are three different open loop process responses.

- Self-regulating – When the loop process variable is stable in MAN mode, and the output is changed, and the process variable comes to rest at a new steady-state, the loop is self-regulating. See figure 16-4-3. FIC-106 and FIC-206 on the PLP are self-regulating.

- Integrating – When controller output is changed in MAN mode, the error continues to increase or decrease at a near constant rate. Examples include: level loops like LIC-101 on the PLP, batch vessel composition, and pressure and temperature loops on crystallizers and evaporators. See figure 16-4-4.

- Runaway – When the controller output is changed in MAN mode, the error increases or decreases at an exponentially increasing rate. Examples include: polymerization reactor temperature control, bio-reactor cell concentration control and axial compressor speed control. See figure 16-4-5.

Figure 16-4-3 Self Regulating Open Loop Response

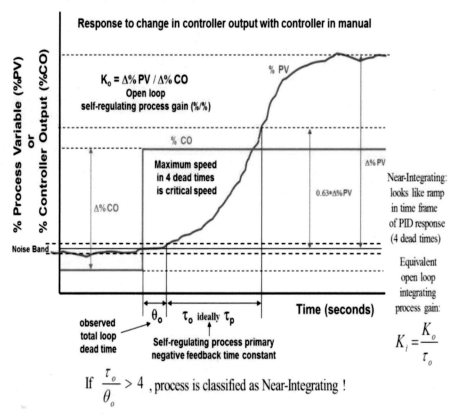

Tuning and Control Loop Performance 4th Edition Momentum Press

Figure 16-4-4 Integrating Open Loop Response

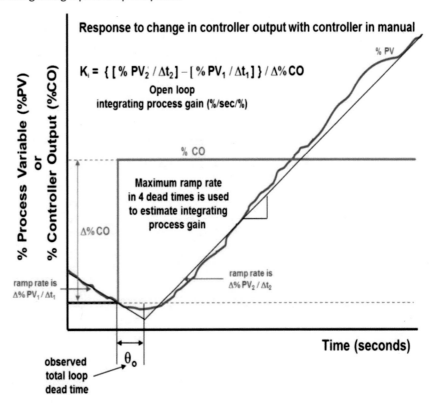

Tuning and Control Loop Performance 4th Edition Momentum Press

Figure 16-4-5 Runaway Open Loop Response

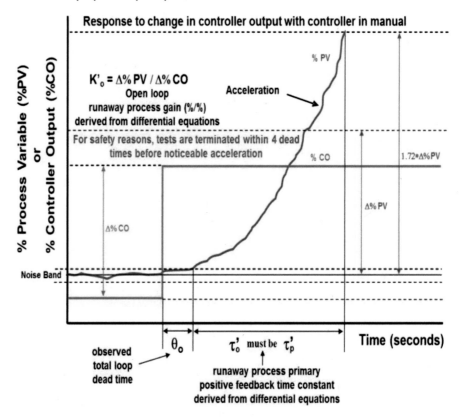

Tuning and Control Loop Performance 4th Edition Momentum Press

As stated above, depending on the application, the control engineer may turn off parts of the control algorithm. Examples include:

- Proportional only controllers – These controllers are primarily used to control level in a surge tank. Proportional only control will propagation of load disturbance to downstream flow controllers in a level-flow cascade scheme. A proportional only controller closed loop response will always produce an offset from setpoint.

- Proportional-Integral controllers – These controllers are primarily used to control flow, tight liquid level applications and pressure applications where some overshoot is acceptable.

- Proportional-Integral-Derivative controllers – Theses controllers are used on reactor, column and in-line temperature and pH applications and pressure control where overshoot is not acceptable.

There are numerous tuning methods available, each with their own equations. They include:

- Lambda for self-regulating processes

- Lambda for integrating processes

- IMC for self-regulating processes

- IMC for integrating processes

- SMIC for self-regulating processes

- SMIC for integrating processes

- Ziegler-Nichols open loop response tuning

- Ziegler-Nichols closed loop response tuning

- SMC for self-regulating processes

- SMC for integrating processes

- SMC for runaway processes

The workshops included in this section explore Ziegler-Nichols open loop and Lambda for self-regulating processes.

16.5   Additional Information

The book *Tuning and Control Loop Performance (Fourth Edition)* by Gregory K. McMillan is an excellent reference for loop tuning and PID control.

16.6   Workshop – Controller Action

Step 16.6.1

From the graphic screen, set the Complexity to Flow Control and Tank Operation to STARTUP. PLP will startup. Allow FIC-106 and FIC-206 to stabilize at around 5.0 GPM. *You may need to run Fill Tank 1 first to get TK-100 above 55%.*

Step 16.6.2

On the main graphic, click on FIC-206 to bring up the faceplate. Place FIC-206 in MAN mode and move the output down 5%. After the loop reaches steady-state, calculate the percent error from setpoint. Is the error negative or positive? Based on the result, is the loop configured as direct or reverse acting?

_____

_____

_____

Step 16.6.3

From the FIC-206 faceplate, click on the Loop Tuning Detail. Look for the PID action. Did it match your results in step 16.6.2?

_____

_____

Step 16.6.4

Repeat steps 16.6.1 through 16.6.3 for LIC-101 but increase the output 5% or 2 GPM.

_____

_____

_____

16.7 Workshop – Open Loop Controller Tuning

Step 16.7.1

From the main graphic Toolbar, launch DeltaV explorer. In the left-hand pane, expand the Physical Network and drill down to the PLP Professional Plus/Continuous Historian. Right click on the Continuous Historian and assign the DUALFLOW area to the Continuous Historian if not already assigned.

Graphic 16-7-1 Adding an Area for Trending

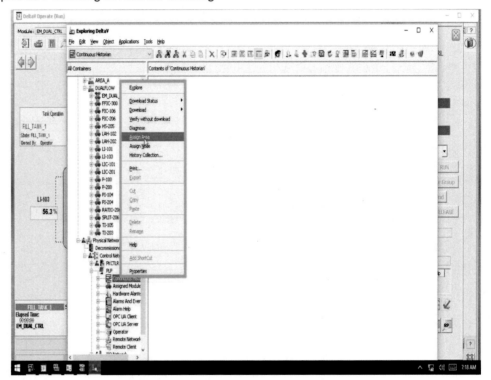

Step 16.7.2

In the left-hand pane, expand the System Configuration and drill down to the FIC-206 module under the DUALFLOW area. Right click on FIC-206 and select History Collection. What appears is that the loop is setup to trend parameters PV, SP and OUT of the PID block. Click on the Modify button and change the sampling rate to 1.0 second on all three parameters. Change the Deviation setting on all three parameters to 0.1 engineering units to increase the resolution of the trend.

Step 16.7.3

In DeltaV Explorer in the left window, right click on FIC-206 and download the module. **Cancel any uploads during the download process.**

Graphic 16-7-3 DeltaV Module Download

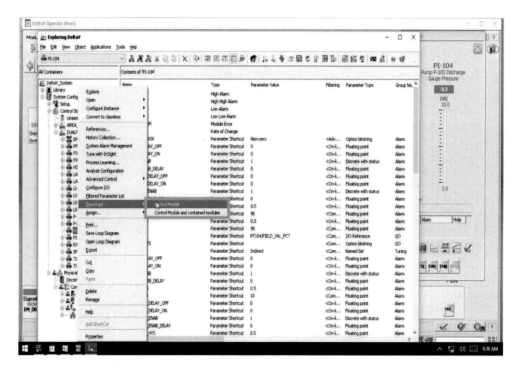

Step 16.7.4

In DeltaV Explorer in the left window, expand the Physical Network and expand the Control Network. Right click on the PLP device and select download ProfessionalPlus Station. **Cancel any uploads during the download proces**s. The Continuous Historian will now begin recording the PV more frequently.

Graphic 16-7-4 DeltaV ProfessionalPlus Download

Step 16.7.5

From the graphic screen, set the Complexity to Flow Control and Tank Operation to STARTUP. PLP will startup. Allow FIC-106 and FIC-206 to stabilize at around 5.0 GPM. *You may need to run Fill Tank 1 first to get TK-100 above 55%.*

Step 16.7.6

From the main graphic, click on FIC-206 to bring up the loop faceplate. From the Icons at the bottom of the faceplate, launch Process History View. Below the tread, right click on the PID1/PV parameter and change the Y scale to 4 to 10 GPM. Right click on the PID1/SP parameter and change the Y scale to 4 to 10 GPM. Using the time scaling functions on the PHV tool bar, set the trend up so the entire width is about 30 seconds. On the trend, click to the right of the trend lines so the trend is updating on real time. On the PHV Toolbar under View/Options and the Chart and Theme Preferences tab, turn on the X and Y grid.

Graphic 16-7-6 PHV Y-Scale Modification

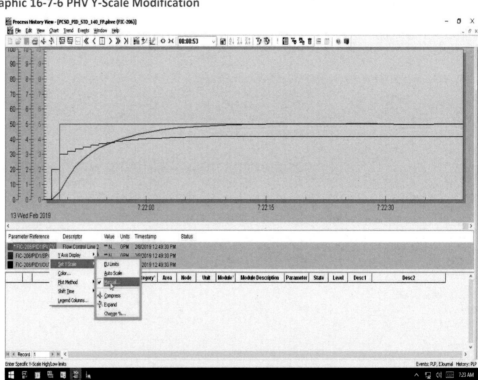

Step 16.7.7

From the main graphic, click on FIC-206 and put it in MAN mode with an output of 40%. Allow the loop to reach steady-state. Once at steady-state, change the output to 50% and allow the loop to reach steady-state again. Do a screen capture of the trend. Copy the trend onto a virus free thumb drive and print it out. Draw a tangent line at the steepest inflection point of the curve as shown in Figure 16-7-7. Calculate the slope R using percent of scale, not engineering units, and L (dead time) in seconds.

R = _____ %/second = $\frac{\Delta\%PV}{\Delta t}$

L = _____ seconds

U (step change in percent) = 10%

Figure 16-7-7 FIC-206 Open Loop Response

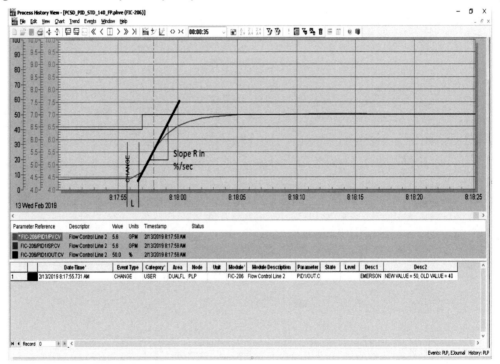

Step 16.7.8

Using the Ziegler-Nichols tuning equations for PI self-regulating loops below, calculate the gain and reset tuning parameters.

$K_c = (0.9) \left(\frac{U}{(L)(R)}\right)$

$T_i = (3.3)(L)$

$T_d = 0$ seconds

$K_c = $ _____

$T_i = $ _____ seconds

Step 16.7.9

From the main graphic, click on FIC-206 to bring up the faceplate. Change the loop mode to AUTO with a setpoint of 5.0 GPM. Allow the loop to stabilize at around 5.0 GPM. At the bottom of the faceplate, click on the Loop Tuning Detail Icon and record the current gain, reset and rate.

$K_c = $ _____

$T_i = $ _____ seconds

$T_d = $ _____ seconds

Step 16.7.10

From the faceplate for FIC 206, set the loop setpoint to 8.0 GPM. Go to the PHV application and screen capture the response when the loop stabilizes around 8.0 GPM. Do a screen capture of the closed loop response with the default tuning settings.

Step 16.7.11

Set FIC-206 setpoint to 5.0 GPM and allow the loop to stabilize. From the FIC-206 Loop Tuning Detail, change the controller gain and reset to the values calculated in step 16.7.8.

Step16.7.12

Repeat step 16.7.10. Do the original or Zeigler-Nichols tuning settings get to setpoint quicker on the first oscillation? Which tuning settings stabilize quicker? Which tuning settings overshoots the setpoint?

_____

_____

_____

Step 16.7.13

Set FIC-206 setpoint to 5.0 GPM and allow the loop to stabilize.

Step 16.7.14

Using the response captured in step 16.7.7 and the definitions on Figure 16-4-3 for self-regulating open loop responses, calculate the Lambda tuning settings per the equations below.

$T_i = \tau_0$ = (open loop time constant based on 63% of the total % PV change)

$T_i =$_____seconds

θ_0 = (open loop dead time) = (time stamp of 10% step change minus the time where the loop first starts to respond to the change outside the noise band)

$\theta_0 =$_____ seconds

$\lambda = (3)(\theta_0)$ = (three times the open loop dead time to minimize nonlinearities)

$\lambda =$_____seconds

$K_o = \frac{\Delta\%PV}{\Delta\%CO}$ at steady-state

$K_c = \frac{T_i}{(K_o)(\lambda+\theta_0)}$

$K_c =$_____

$T_d = 0$ seconds

Step 16.7.15

From the FIC-206 Loop Tuning Detail, change the controller gain and reset to the values calculated in step 16.7.14.

Step 16.7.16

Repeat step 16.7.10. Do the original or Lambda tuning settings get to setpoint quicker on the first oscillation? Which tuning settings stabilize quicker? Which tuning settings overshoots the setpoint? Is the Lambda or Ziegler-Nichols tuning better?

_____

_____

_____

Step 16.7.17

From the FIC-206 Loop Tuning Detail, change the controller gain and reset to the original values captured in step 16.7.9.

16.8   Conclusions

Loop tuning is critical to an operating facility to produce in-spec product and to reduce costs by minimizing energy usage. A small improvement in performance can have a dramatic effect on cost and cycle time. There are several methods available to tune a loop, but all require an understanding of the process constraints. The tuning settings calculated may or may not meet the objectives associated with the loop's function.

Not all manufacturers use the same PID algorithm and in many cases the time units are different. Before attempting to tune a loop in a production environment, make sure operations is aware of the job and always have a firm understanding of the algorithm the control system uses.

For more information on Emerson training classes visit the following website:

https://www.emerson.com/en-us/automation/services-consulting/educational-services

Section 17: Cascade Control

17.1 Objectives

When the student has completed this module, the student will:

- Understand why cascade control is used.

- Understand and be able to describe the requirements for effective cascade control.

- Be able to tune a cascade loop.

- Understand why signal propagation is so important for good cascade control.

17.2 Intended Audience

Instrument Technicians

Process Engineers

17.3 Prerequisites

Section 2 PLP Infrastructure

Section 3 PLP Instrumentation

Section 4 DeltaV Navigation

Section 5 DeltaV Modes and Signal Status

Section 16 PID Loop Tuning

17.4 Discussion

Cascade control is implemented by sending the output of an upper (master) PID loop to the setpoint of a lower (slave) PID loop. This discussion will focus on two level cascades schemes, but if you consider a valve positioner, many times there are three loops cascaded together. The advantages of cascade control include:

- Minimizing the effects of disturbances on the upper loop by filtering out disturbances in the lower loop.

- Isolating nonlinearities from the upper loop by enforcing limits on the lower loop's control range.

- Better response to disturbances because the closed loop response time of the upper loop is smaller than if the primary variable would have been setup as a single loop.

Figure 17-4-1 shows how a Cascade control scheme would be setup in DeltaV.

Figure 17-4-1 DeltaV Cascade Control

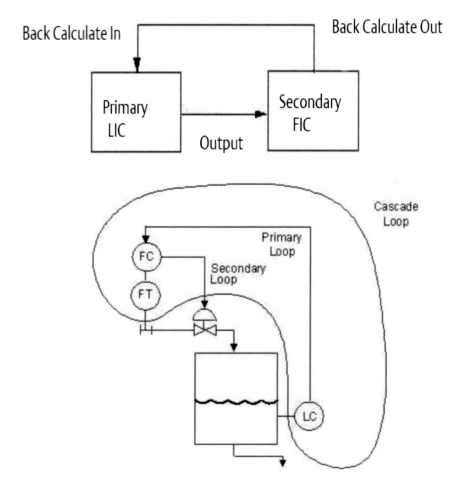

There are several rules to consider when it comes to deploying cascade control effectively.

- The upper loop process variable and the lower loop process variable must be interrelated. For example, in a level to flow cascade scheme, the level in a tank (upper loop PV) is dependent on the flow rate out of the tank (lower loop PV).

- The closed loop response of the lower loop must be at least 5 times faster than the closed loop response of the upper loop. Failure to maintain this rule can result in larger peak and integrated error on the primary loop than if it was setup in direct control of the final control element. If the 5 times faster response is not inherent to the process, the upper loop could be de-tuned to slow its response.

- The output of the upper loop should be configured in the same units and span as the lower loop process variable.

- Signal status (health) and setpoint limits of the lower loop must be propagated up to the upper loop. For example, if a lower flow loop is setup with a span of 0 to 100 GPM, but with setpoint limits of 30 to 70 GPM, the upper loop needs to know if the process is at the lower loop's setpoint limits so it can stop trying to integrate the error and not windup. Another example; if the process variable or the output of the lower loop are bad, the upper loop should respond to this by minimizing the integral action since it essentially lost the ability to control.

- Tune the lower loop first, then with the lower loop in CAS mode, tune the upper loop. Do this so the effect of the lower loop's dead time is included in the upper loop's open loop response.

- When the lower loop is not in CAS mode, the output of the upper loop should track the setpoint of the lower loop to allow for smooth transfer when the lower loop is put back in CAS mode.

17.5   Additional Information

The book *Tuning and Control Loop Performance (Fourth Edition)* by Gregory K. McMillan is an excellent reference for cascade control.

17.6 Workshop – Cascade Control Response

Step 17.6.1

From the main graphic, click on LIC-101 to bring up the faceplate. On the bottom of the faceplate, click on the Loop Tuning Detail and record the current gain, reset and rate.

K_c = _____

T_i = _____ seconds

T_d = _____ seconds

Step 17.6.2

From the main graphic Toolbar, launch DeltaV Explorer. In the left-hand pane, expand the System Configuration and drill down to the LIC-101 module under the DUALFLOW area. Right click on the module and select History Collection. What appears is that the loop is setup to trend parameters PV, SP and OUT of the PID block. Click on the Modify button and change the sampling rate to 1.0 seconds on all three parameters. Change the Deviation setting on all three parameters to 0.1 engineering units to increase the resolution of the trend.

Step 17.6.3

In DeltaV Explorer in the left window, right click on LIC-101 and download the module. **Cancel any uploads during the download process.**

Step 17.6.4

In DeltaV Explorer in the left window, expand the Physical Network and expand the Control Network. Right click on the PLP device and select download ProfessionalPlus Station. **Cancel any uploads during the download process**. The Continuous Historian will now begin recording the PV status.

Step 17.6.5

From the graphic screen, set the Complexity to Flow Control and Tank Operation to STARTUP. PLP will startup. Allow FIC-106 and FIC-206 to stabilize at around 5.0 GPM and LIC-101 to stabilize at around 45%. *You may need to run Fill Tank 1 first to get TK-100 above 55%.*

Step 17.6.6

From the main graphic, click on LIC-101 to bring up the loop faceplate. From the Icons at the bottom of the faceplate, launch Process History View. Below the tread, right click on the PID1/PV parameter and change the Y scale to 40 to 55 %. Right click on the PID1/SP parameter and change the Y scale to 40 to 55%. Using the time scaling functions on the PHV tool bar, set the trend up so the entire width is about 20 minutes. On the PHV Toolbar under View/Options and the Chart and Theme Preferences tab, turn on the X and Y grid.

Step 17.6.7

Click on LIC-101 and place the loop in AUTO mode and move the setpoint up 5%. Go back to the trend setup in step17.6.6. When LIC-101 stabilizes, do a screen capture of the trend. Copy the trend onto a virus free thumb drive and print it out. Is the system oscillatory? What does it suggest about the default tuning?

_____

_____

_____

Step 17.6.8

Shut down the PLP.

Step 17.6.9

From the graphic screen, set the Complexity to Flow Control and Tank Operation to STARTUP. PLP will startup. Allow FIC-106 and FIC-206 to stabilize at around 5.0 GPM and LIC-101 to stabilize at around 45%. *You may need to run Fill Tank 1 first to get TK-100 above 55%.* When LIC-101 is stable at setpoint, change FIC-106 and FIC-206 to AUTO with setpoints of 7.0 GPM and again wait for the level to stabilize.

Step 17.6.10

Assuming FIC-206 tuning is acceptable and with LIC-101 stable, do an open loop response to find LIC-101 Lambda tuning values by quickly placing FIC-106 in CAS mode, placing LIC-101 in MAN mode and decreasing the output by 2 GPM or 5% of span. Record the trend.

Step 17.6.11

Using the open loop response in step 17.6.10, the equations below and the information in Figure 16-4-4, calculate the Lambda tuning settings of LIC-101. Assume rate of change of PV_1 is zero.

θ_0 = open loop dead time

$\lambda = (3)(\theta_0)$ = (three times the open loop dead time to minimize nonlinearities)

$$K_i = \frac{\left(\frac{\Delta\%PV_2}{\Delta t}\right) - \left(\frac{\Delta\%PV_1}{\Delta t}\right)}{\Delta\%CO} \qquad \text{where } \left(\frac{\Delta\%PV_1}{\Delta t}\right) = 0$$

$T_i = (2)(\lambda) + \theta_0$

$$K_c = \frac{T_i}{(K_i)(\lambda + \theta_0)^2}$$

$(0.5)(\theta_0) < T_d < (0.25)(T_i)$

$\theta_0 =$ _____ seconds

$\lambda =$ _____ seconds

$K_i =$ _____ %/sec/%

$K_c =$ _____

$T_i =$ _____ seconds

$T_d =$ _____ seconds

Step 17.6.12

Place FIC-206 in AUTO mode with a setpoint of 5.0 GPM, FIC-106 in CAS mode and LIC-101 in AUTO mode with a setpoint of 45.0%. Once stable, go to the LIC-101 Loop Tune Detail and enter in the Lambda tuning settings calculated in step 17.6.11.

Step 17.6.13

Move the setpoint of LIC-101 up 5.0% and wait for the system to stabilize. Do a screen capture of the closed loop response with the Lambda tuning settings. Compare the cascade closed loop response with the default tuning to the cascade closed loop response with Lambda tuning and record observations.

_____

_____

_____

Step 17.6.14

From the main graphic Toolbar, launch DeltaV explorer. In the left-hand pane, expand the system configuration and drill down to the FIC-106 module under the DUALFLOW area. Click on FIC-106 and the parameters and function blocks which makeup FIC-106 appear in the right pane. Double click the PID1$CONTROL_OPTS parameter, unclick "Use default library values" and under Properties scroll down and turn on Bypass Enabled. Once downloaded and Bypass is turned on, this will bypass the FIC-106 loop and LIC-101 will be directly in control of FV-106.

Step 17.6.15

In DeltaV Explorer in the left window, right click on LIC-101 and download the module. **Cancel any uploads during the download process.**

Step 17.6.16

Change the tuning settings of LIC-101 to the default tuning settings recorded in step 17.6.1. Go to the FIC-106 Loop Tune Detail and change the HI OUT LIM parameter from 20 to 25 to prevent interfering with the open loop response.

Step 17.6.17

From the graphic screen, set the Complexity to Flow Control and Tank Operation to STARTUP. PLP will startup. Allow FIC-106 and FIC-206 to stabilize at around 5.0 GPM and LIC-101 to stabilize at around 45%. *You may need to run Fill Tank 1 first to get TK-100 above 55%.*

Step 17.6.18

Place FIC-106 in MAN mode so it will accept a Bypass command.

Step 17.6.19

From DeltaV Explorer, expand the System Configuration and drill down to the FIC-106 module under the DUALFLOW area. Right click on the module and select Open with Control Studio On-Line Debug Mode. In the right pane, find the PID1 block and click on it. In the lower left pane, scroll down to the Bypass parameter and set it to True. LIC-101 is now in direct control of FV-106.

Step 17.6.20

Place FIC-206 in AUTO with a setpoint of 7.0 GPM. Place FIC-106 in CAS mode. Place LIC-101 in MAN mode and adjust the output so that the PV of FIC-106 is 7.0 GPM. This will cause the level in LIC-101 to stay constant and the PV_1 rate of change for the Lambda open loop test will be zero.

Step 17.6.21

Run the open loop response with the cascade scheme bypassed by decreasing the output of LIC-101 by 2.0 GPM or 5.0% of span. Record the trend.

Step 17.6.22

Using the open loop response in step 17.6.21, the equations below and the information in Figure 16-4-4, calculate the Lambda tuning settings of LIC-101 with the cascade scheme bypassed. Assume rate of change of PV_1 is zero.

θ_0 = open loop dead time

$\lambda = (3)(\theta_0)$ = (three times the open loop dead time to minimize nonlinearities)

$K_i = \dfrac{\left(\frac{\Delta\%PV_2}{\Delta t}\right) - \left(\frac{\Delta\%PV_1}{\Delta t}\right)}{\Delta\%CO}$ where $\left(\frac{\Delta\%PV_1}{\Delta t}\right) = 0$

$T_i = (2)(\lambda) + \theta_0$

$K_c = \dfrac{T_i}{(K_i)(\lambda + \theta_0)^2}$

$(0.5)(\theta_0) < T_d < (0.25)(T_i)$

$\theta_0 =$ _____ seconds

$\lambda =$ _____ seconds

$K_i =$ _____ %/sec/%

$K_c =$ _____

$T_i =$ _____ seconds

$T_d =$ _____ seconds

Step 17.6.23

Enter in the new tuning settings from step 17.7.22 on LIC-101 on the Loop Tuning Detail. Place FIC-206 in AUTO mode with a setpoint of 5.0 GPM, FIC-106 in CAS mode and LIC-101 in AUTO mode with a setpoint of 45.0%. Allow the system to stabilize.

Step 17.6.24

Click on LIC-101 and place the loop in AUTO mode and move the setpoint up 5%. When LIC-101 stabilizes, do a screen capture of the trend. Copy the trend onto a virus free thumb drive and print it out. This is the non-cascade closed loop response. How did it compare with the response captured in step 17.6.13? Was the cascade control better?

_____

_____

_____

Step 17.6.25

Set FIC-106 to MAN mode. From DeltaV Explorer, expand the System Configuration and drill down to the FIC-106 module under the DUALFLOW area. Right click on the module and select Open with Control Studio On-Line Debug Mode. In the right pane, find the PID1 block and click on it. In the lower left pane, scroll down to the Bypass parameter and set it to False.

Step 17.6.26

Change the tuning settings of LIC-101 to the default tuning settings recorded in step 17.6.1. Go to the FIC-106 Loop Tune Detail and restore the HI OUT LIM parameter to 20.

Step 17.6.27

From the main graphic Toolbar, launch DeltaV explorer. In the left-hand pane, expand the system configuration and drill down to the FIC-106 module under the DUALFLOW area. Click on FIC-106 and the parameters and function blocks which makeup FIC-106 appear in the right pane. Double click the PID1$CONTROL_OPTS parameter and under Properties scroll down and turn off Bypass Enabled.

17.7   Workshop – Cascade Loop Signal Status Propagation

Step 17.7.1

Bypass interlocks on P-100.

Step 17.7.2

From the graphic screen, set the Complexity to Flow Control and Tank Operation to STARTUP. PLP will startup. Allow FIC-106 and FIC-206 to stabilize at around 5.0 GPM and LIC-101 to stabilize at 45%. *You may need to run Fill Tank 1 first to get TK-100 above 55%.*

Step 17.7.3

From the graphic screen, click on FIC-106 to bring up the loop faceplate.   Launch Control Studio On-Line and find the PID1 block in the left pane and in the lower right pane, record the status of the block PID1/BKCAL_OUT.

Graphic 17-7-3 FIC-106 PID1 BKCALC_ OUT Status

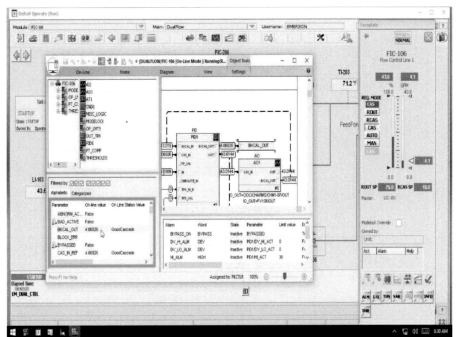

FIC-106 PID1/BKCAL_OUT Status = _____

Step 17.7.4

From the graphic screen, click on LIC-101 to bring up the loop faceplate. Launch Control Studio On-Line and find the PID1 block in the left pane and in the lower right pane, record the status of the PID1 block BKCAL_IN.

Graphic 17-7-4 LIC-101 BKCAL_IN Status

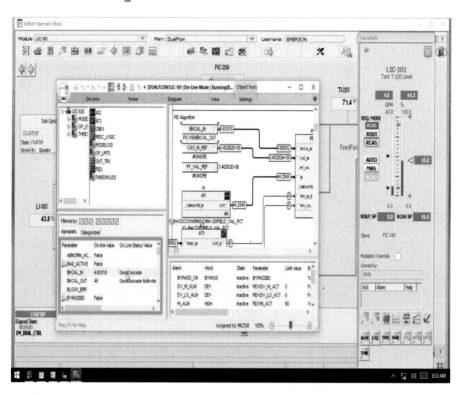

LIC-101 PID1/BKCAL_IN Status = _____

Step 17.7.5

Launch DeltaV Explorer and drill down in the Physical Network to CHM1-6 and FT-106 and launch
Service Tools.

Graphic 17-7-5 DeltaV Explorer AMS Access

Step 17.7.6

Simulate a low flow condition on FT-106 by using the Loop Test simulate function in AMS and doing an Other simulation of 5 mA.

Graphic 17-7-6 AMS Loop Test

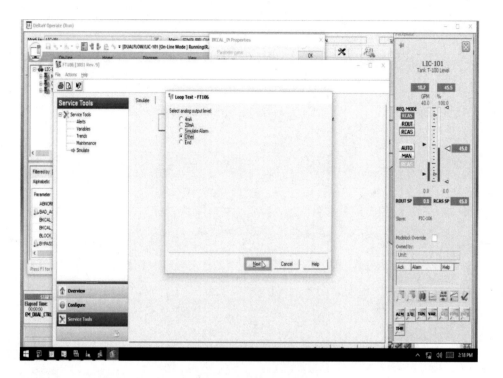

Step 17.7.7

FIC-106 should try and open FV-106 to maintain flow. After FV-106 reaches 100%, open Control Studio On-Line on FIC-106 and record the status of the PID1/BKCAL_OUT parameter.   Does the status reflect the valve being completely open?

FIC-106 PID1/BKCAL_OUT status = _____

_____

_____

Step 17.7.8

Using Control Studio On-Line on LIC-101, record the status of the PID1 block BKCAL_IN. Did the status propagate up to the Master loop?

LIC-101 PID1/BKCAL_IN status = _____

_____

_____

Step 17.7.9

Go back to AMS and remove the simulation on FIC-106.

Graphic 17-7-9 AMS Loop Test Termination

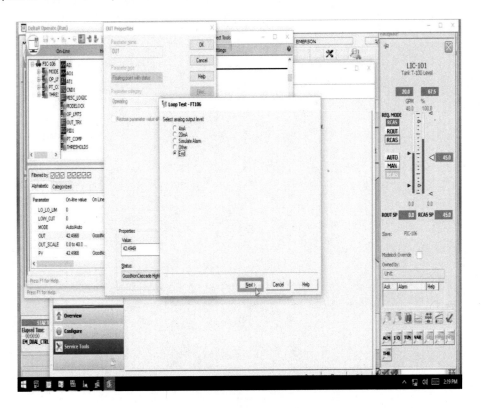

Step 17.7.10

Reinitiate interlocks on P-100.

17.8   Conclusions

Cascade control can improve plant operations by minimizing disturbances on primary control variables. If tuned correctly, it will allow the primary variable to reach setpoint much quicker than if it was in direct control of the final control element.

For cascade control to work properly, the closed loop response time of the lower loop must be at least 5 times faster than the upper loop's response time. The configuration of a master cascade loop must constantly monitor the status of the slave loop and take appropriate action if the slave loop is limited in some fashion.

For more information on Emerson training classes visit the following website:

https://www.emerson.com/en-us/automation/services-consulting/educational-services

Section 18: Split-Range Control

18.1 Objectives

When the student has completed this module, the student will:

- Understand and be able to name the three types of split-range controllers.

- Understand why split range may be implemented with a loop and two manual loaders.

- Understand the importance of signal status when configuring a split-range controller.

- Understand why some split range controllers are implemented with dead band.

18.2 Intended Audience

Instrument Technicians

Process Engineers

Operators

18.3 Prerequisites

Section 2 PLP Infrastructure

Section 3 PLP Instrumentation

Section 4 DeltaV Navigation

Section 10 Control Valve Selection

Section 16 PID Loop Tuning

Section 17 Cascade Control

18.4   Discussion

Split-Range control occurs when one PID controller controls two final control elements. Figure 18-4-1 shows how split-range control is setup in DeltaV. The Input Array and Output Array on the splitter function block determines how the PID output is passed to the two final control elements. One observation that can be made by examining figure 18-4-1 is that the status of each final control element is propagated up to the PID loop via the Back Calculate parameter. This prevents the loop from winding up if there is a problem with either one of the final control elements.

Figure 18-4-1 DeltaV Split Range Controller

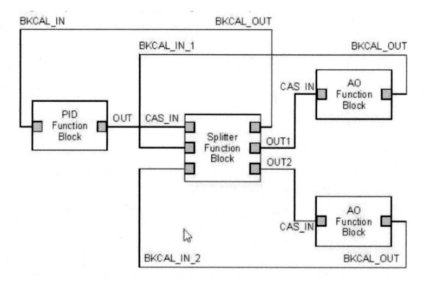

Sometimes split-range control is implemented with two manual loader loops in CAS mode connected to the two outputs of the splitter function block. This is done for maintenance purposes. Each manual loader can be independently manipulated by simply putting the manual loader in MAN mode. This gives maintenance personnel the ability to work on each final control element without needing to understand the split-range scheme.

18.4.1   Exclusive Split-Range Control

One very common split-range controller application is the heating and cooling of a process vessel jacket. One final control element (valve) controls the heating rate, and one final control element (valve) controls the cooling rate. Remember, for optimal PID control, the response must be linear, and the tuning settings need to work over the whole output range. This implies that the open loop process gain for both valves should be roughly the same. To accomplish this, the valves may have to be different sizes. This type of split-range control is an example of exclusive split-range control. Only one final control element is affecting the process at a time. If the controller is at 50%, neither final control element is manipulating the process variable. Exclusive split-range control is typically implemented with a dead band of 1 to 2 percent to prevent disturbances from being introduced when the loop is near setpoint. See figure 18-4-1-1. Exclusive split-range control is also used on pH systems where one valve controls the caustic addition rate and one valve controls the acid addition rate.

Figure 18-4-1-1 Exclusive Split Range Controller Output

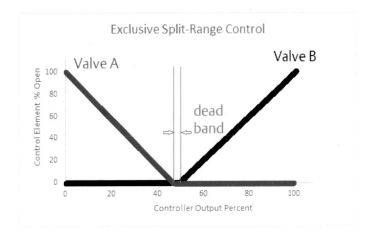

18.4.2   Expanded Split-Range Control

Expanded split-range control is used to expand the range of a final control element. The function of each final control element (valve) is the same. A smaller valve is opened during the first 50% of the PID output. The second larger valve is opened during the last 50% of the PID output. When the controller output is at 100%, both valves are fully open as shown in figure 18-4-2-1. This split-range controller could be used in situations where you have a large intermittent process upset like waste water from a batch system that needs to be neutralized. Most of the time the smaller valve is in control of the reagent, but if needed, the larger valve could open increasing the controller rangeability.

Figure 18-4-2-1 Expanded Split-Range Controller Output

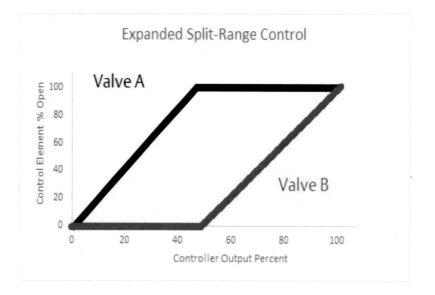

18.4.3   Proportional Split-Range Control

Proportional split-range control is used to ratio two different process streams. In this scheme, the two final control elements are never closed at the same time. As one valve begins to open, the other valve begins to close. Both valves are at 50% open when the PID output is 50% as shown in Figure 18-4-3-1. Proportional split-range has been used for color control to adjust the pigment and base addition rates or for reactor pressure control to adjust feed and product discharge rates.

Figure 18-4-3-1 Proportional Split-Range Controller Output

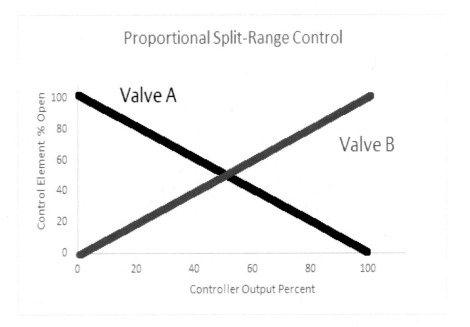

18.5 Workshop – Split-Range Control

Step 18.5.1

From the main graphic, setup Flow Control Type to Split Range.

Figure 18-5-1 Split Range Control Setup

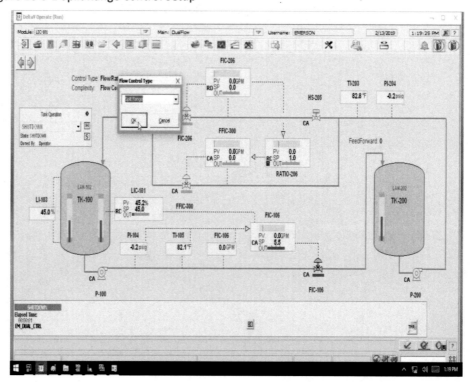

Step 18.5.2

From the main graphic Toolbar, launch DeltaV Explorer. Expand the System Configuration and drill down to the SPLIT-206 module in the DUALFLOW area. In the right pane, right click on the SPLTR1$OUT parameter and select properties.

Graphic 18-5-2 Split-Range Parameter Access

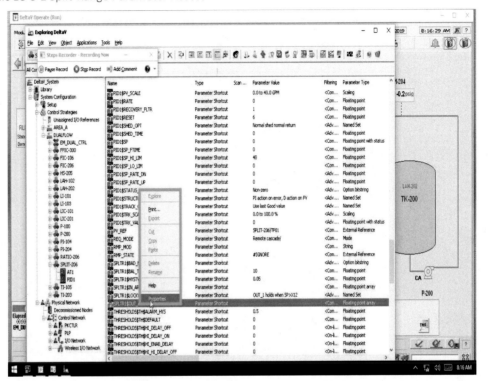

Step 18.5.3

Unclick the "Use default value from library values" and record the property values of 1, 2, 3 and 4. Modify Properties values 1 and 2. Set 1 to 100 and set 2 to 0. Press OK when complete.

Graphic 18-5-3 Splitter Function Block Modification

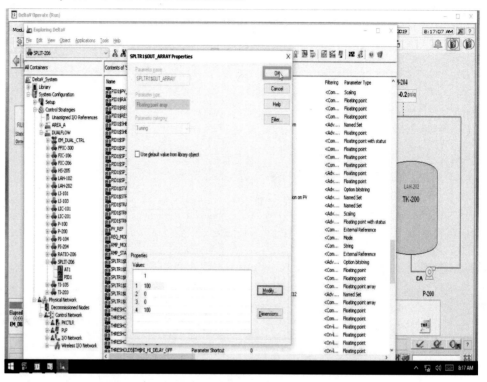

Step 18.5.4

In DeltaV Explorer in the left window, right click on SPLIT-206 module and download the module. **Cancel any uploads during the download process.**

Step 18.5.5

From the main graphic, click on the SPLIT-206 module to bring up the faceplate. Place SPLIT-206 in MAN mode. Complete the table below by increasing the output 5% for each reading and recording the valve position of FV-206 and FV-300.

Table 18-5-5 SPLIT-206 Split Range Response

SPLIT-206 % Output	FV-206 % Open	FV-300 % Open
0		
5		
10		
15		
20		
25		
30		
35		
40		
45		
50		
55		
60		
65		
70		
75		
80		
85		
90		
95		
100		

Step 18.5.6

What type of split-range controller is it? What could this type of split-range controller be used for?

_____

_____

_____

Step 18.5.7

Restore the property values recorded in step 18.5.3.

Step 18.5.8

In DeltaV Explorer in the left window, right click on the SPLIT-206 module and download the module. **Cancel any uploads during the download process.**

18.6   Conclusions

Split-Range control is used extensively in the process industry on PID controllers to control two outputs related to one another. There are three ways to setup split-range control: exclusive, expanded and proportional.

To optimize a split-range controller, the final control elements need to be sized properly to produce a linear response across the whole output range.

For more information on Emerson training classes visit the following website:

https://www.emerson.com/en-us/automation/services-consulting/educational-services

Section 19:   Answer Key

19.1   Lab Answers Key Note

All of these labs except for section 17 were developed to be run on a PLP unit with a single flow path. If doing the labs on a dual path PLP unit, simply set module FFIC-300 Out Hi Lim parameter on the Loop Tune Detail to 0%.

19.2   Section 2.6 Answers – PLP P&ID Completion

Step 2.6.2

Figure 2-6-2 PLP Completed P&ID

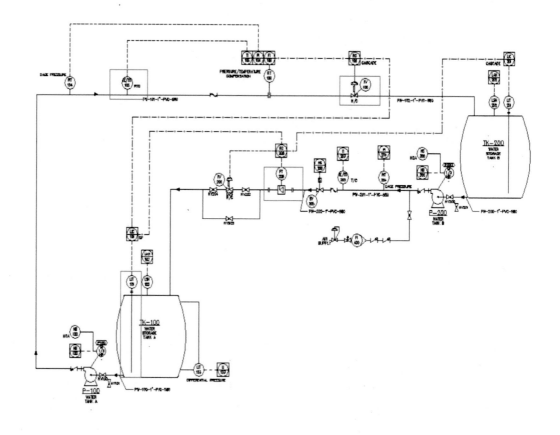

Step 2.6.3

Figure 2-6-3 PLP P&ID Flow Paths

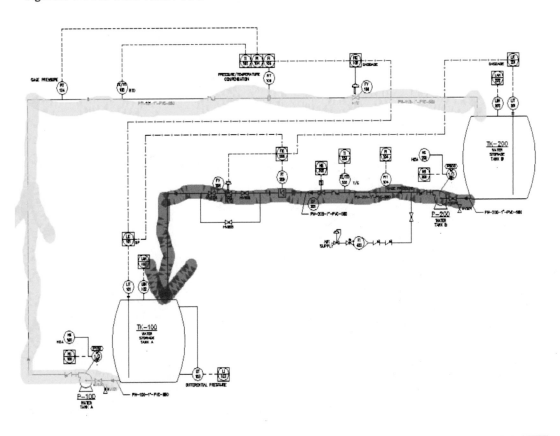

19.3 Section 3.5 Answers – Instrument Sizing and Materials of Construction Identification

Step 3.5.3

Model: 3051TG1A2B21A5B4D4M5

3051T = Pressure Transmitter Pipe Mount

Measurement

G = Gage Pressure

Size

1 = 0 -30 PSI

Transmitter Output

A = 4–20 mA with Digital Signal Based on HART Protocol

Process Connection Style

2B = 1/2–14 NPT Female

Isolation Diaphragm

2 = 316L SST

Sensor Fill Fluid

2 = Inert (Fluorinert® FC-43)

Housing Material

A = Polyurethane-covered Aluminum

Integral Assembly

S5 = Assemble to Rosemount 306 Integral Manifold

Mounting Bracket

B4 = Bracket for 2-in. pipe or panel mounting, all SST

External Calibration

D4 = Zero and Span Push Buttons

Display Type

M5 = LCD display

19.4   Section 3.6 Answers– Instrument Calibration and Maintenance

Step 3.6.1

Instrument Verification Procedure

1. Obtain required line break, LOTO (Lock Out Tag Out) and hot work permits.

2. Put on the proper PPE (Personal Protective Equipment) as outlined in the permit.

3. Verify that the NIST pressure calibration standard is still in calibration as noted on the calibration due date sticker.

4. Close pressure isolation valve on PIT-104 and lock it out.

5. Open the drain plug on PIT-104 located above the isolation valve. Make sure you are clear of the direction of the vent.

6. Slowly relieve pressure on PIT-104 by slowly opening the vent valve located above the isolation valve.

7. Flush out the port with a compatible process flush fluid.

8. With PIT-104 open to atmosphere, record the pressure reading on the transmitter and on PIT-104 on the operating console. Verify the zero-pressure reading is within the specified tolerance and alert quality and supervision as required by the company's policy.

9. Tube up the pressure calibration standard and a hand pump in parallel with the PIT-104 drain port. Pump up the pressure until you are at mid-range on the PIT-104 scale as noted on the PIT-104 faceplate on DeltaV. Record the pressure reading on the transmitter and on PIT-104 on the operating console. Verify the pressure reading is within the specified tolerance and alert quality and supervision as required by the company's policy.

10. Pump up the pressure until you are at full-range on the PIT-104 scale as noted on the PIT-104 faceplate on DeltaV. Record the pressure reading on the transmitter and on PIT-104 on the operating console. Verify the pressure reading is within the specified tolerance and alert quality and supervision as required by the company's policy.

11. Vent PIT-104 to atmosphere.

12. Assuming PIT-104 was in calibration, reinstall the drain plug.

13. Close the PIT-104 vent valve.

14. Remove the lock on the PIT-104 isolation valve and open it.

15. Sign all of the permits and return the permits to production supervision and alert them the work is complete.

16. Submit calibration paperwork to quality and maintenance as required by the company's policy.

Step 3.6.2

Instrument Removal Procedure

1. Obtain required line break, LOTO (Lock Out Tag Out) and hot work permits.

2. Put on the proper PPE (Personal Protective Equipment) as outlined in the permit.

3. Close pressure isolation valve on PIT-104 and lock it out.

4. Open the drain plug on PIT-104 located above the isolation valve.

5. Slowly relieve pressure on PIT-104 by slowly opening the vent valve located above the isolation valve.

6. Remove charm CHM1-05 to isolate power to PIT-104.

7. Document wiring numbers and terminals on PIT-104 for re-installation.

8. Un-wire PIT-104.

9. Flush out the port with a compatible process flush fluid.

10. Un-screw PIT-104 off the isolation manifold.

11. Assuming the transmitter is going to be left out of the line, sign all of the permits and return the permits to production supervision and alert them the work is complete.

19.5   Section 4.8 Answers – Using Control Studio On-Line

Step 4.8.4

The diagnostic indicates an Input Error and the faceplate PV background color changed. A Red X appears next to the AI block indicating an issue with the status of the signal.

Graphic 4-8-4 DeltaV AI1 Status

Step 4.8.5

Status of the OUT parameter became Bad NoCommNUV LowLimited.

Graphic 4-8-5 DeltaV AI1 Value and Status

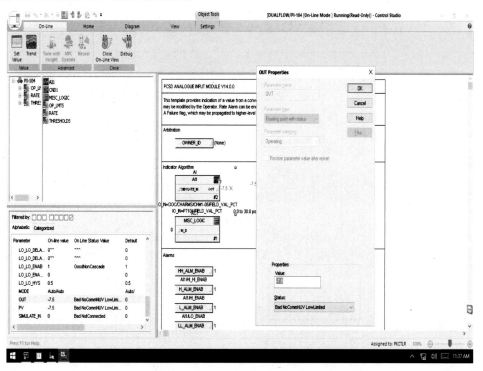

19.6 Section 4.9 Answers – Using the DeltaV Continuous Historian

Step 4.9.12

PV goes from 0.0 PSIG to -7.5 PSIG. Status goes from 128 (GoodNonCascade NonSpecific Not Limited) to 13 (BadDeviceFailureLowLimited).

Graphic 4-9-12 DeltaV Status Trend

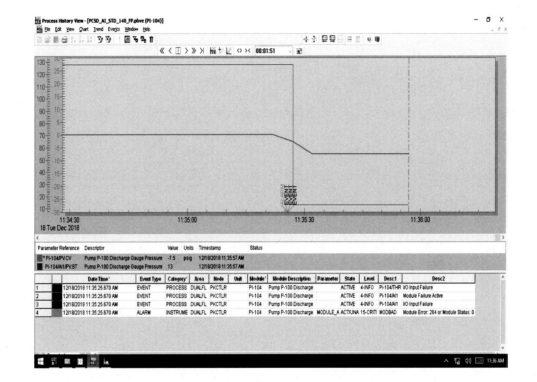

19.7   Section 4.10 Answers – Using DeltaV Diagnostics

Step 4.10.4

Prior to Charm removal, PV = 0.0% and status is Good.

Graphic 4-10-4 DeltaV Charm Status

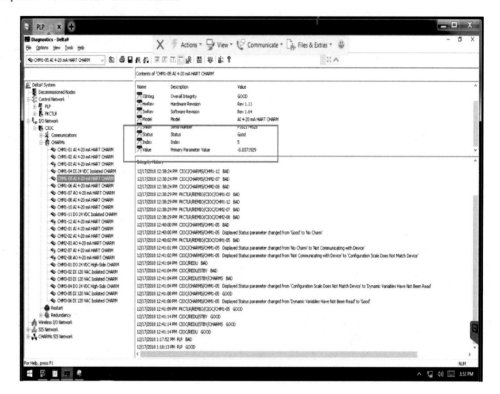

Step 4.10.6

After Charm removal, status goes to No Charm.

Graphic 4-10-6 DeltaV Charm Status

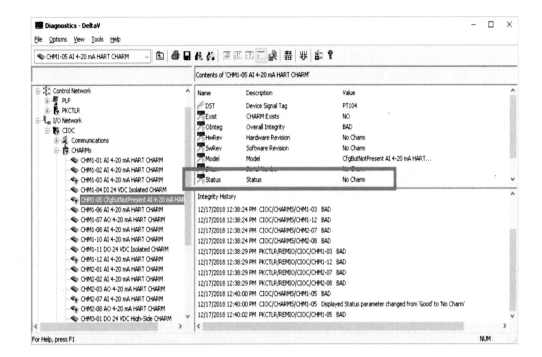

19.8 Section 5.5 Answers – Effect of Mode on Control

Step 5.5.4

Valve remains in last position and the flow (PV) goes down. The loop setpoint tracks the PV.

Graphic 5-5-4 DeltaV Manual Response

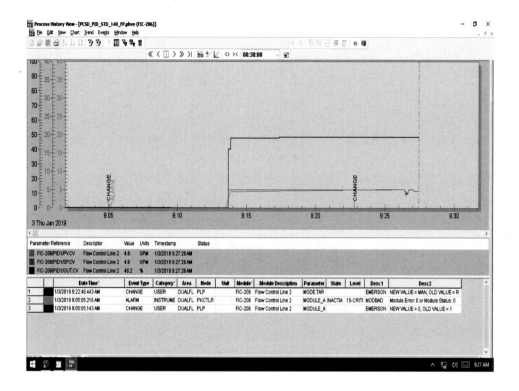

Step 5.5.5

Loop begins to control and increases output to maintain the 5.0 GPM setpoint. At the time the loop was put into AUTO, the setpoint matched the PV to achieve bumpless transfer. This is important in an operating facility to prevent process upsets resulting in potential quality and safety issues. This will give the operator the ability to slowly increase the setpoint over time reducing process variability.

Graphic 5-5-5 DeltaV AUTO Response

Step 5.5.6

The setpoint remained at 5.0 GPM because that is what the sequencer had set the RCAS_IN value to. When the manual valve is pinched back, the loop increases the output to maintain 5.0 GPM.

Graphic 5-5-6 DeltaV RCAS Response

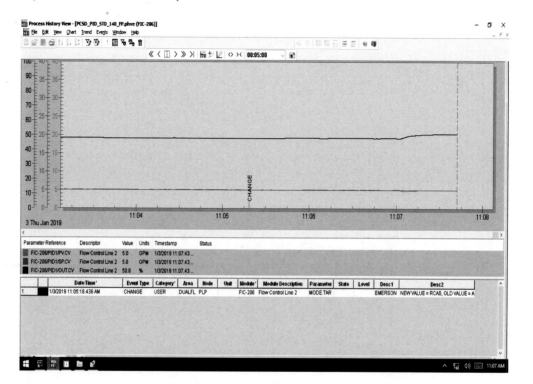

Step 5.5.7

FIC-206 output goes to 20% which was the ROUT SP on the FIC-206 faceplate. The loop setpoint tracks the PV.

Graphic 5-5-7 DeltaV ROUT Response

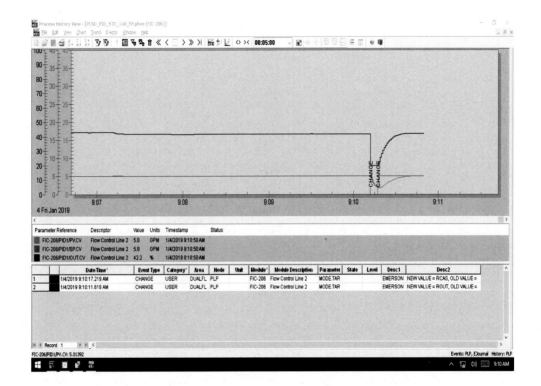

Step 5.5.9

The level will initially go up in TK-101. This will cascade down to loop FIC-106, and the setpoint of FIC-106 will increase to bring the level back to the setpoint. After a while FIC-106 should stabilize at around 4.5 GPM.

Graphic 5-5-9 DeltaV CAS Response

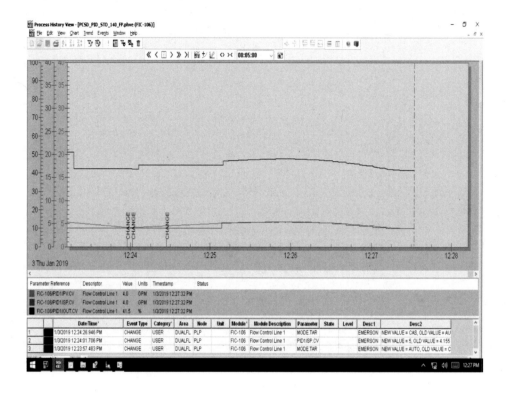

Step 5.5.11

The actual mode of LIC-101 goes to IMAN since it lost control and the output begins tracking the setpoint of FIC-106.

Graphic 5-5-11 DeltaV IMAN Result

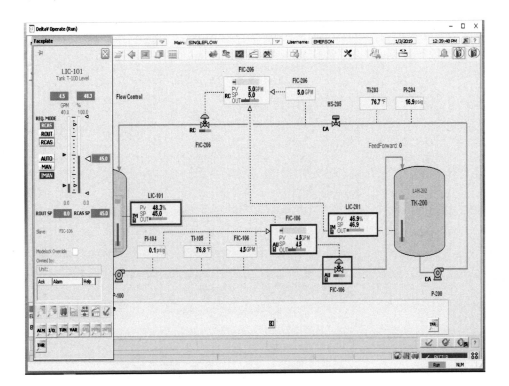

Step 5.5.14

The Actual mode of FIC-106 goes to LO due to tracking condition #2 shown on the interlock detail of FIC-106. The output of FIC-106 goes to 0% which is the TRK_IN value. The unit shuts down.

Graphic 5-5-14 DeltaV LO Result

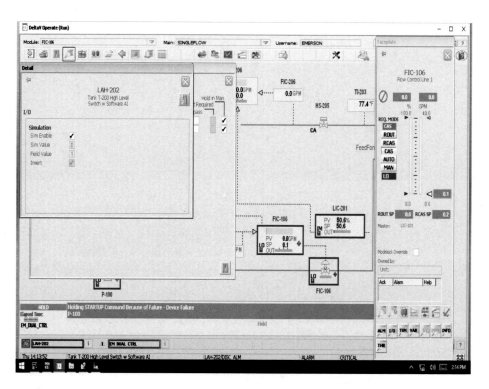

19.9   Section 5.6 Answers – Propagation of Signal Status

Step 5.6.1

GoodNonCascade

Step 5.6.2

GoodCascade

Step 5.6.5

GoodNonCascade HighLimited. The HighLimited represents the saturated transmitter state.

Graphic 5-6-6 Input Status Result

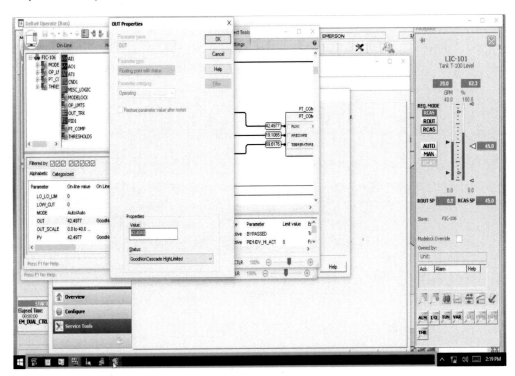

Step 5.6.6

GoodCascade LowLimited. The status was propagated up to the Master loop via the BKCAL_IN parameter.

Graphic 5-6-7 DeltaV PID Status

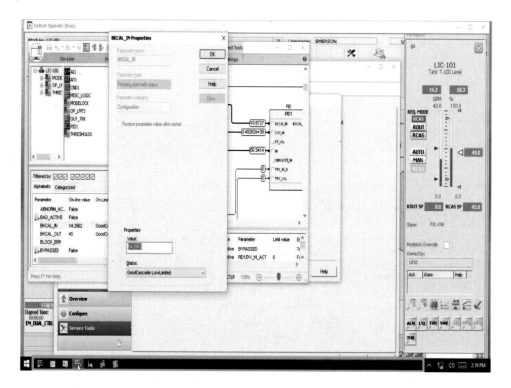

19.10 Section 6.5 Answers – Determining Controller Loading and Free Memory

Step 6.5.2

Free Memory = 715681748 bytes

Free ProcessorTime = ???

Graphic 6-5-2 DeltaV Controller Status

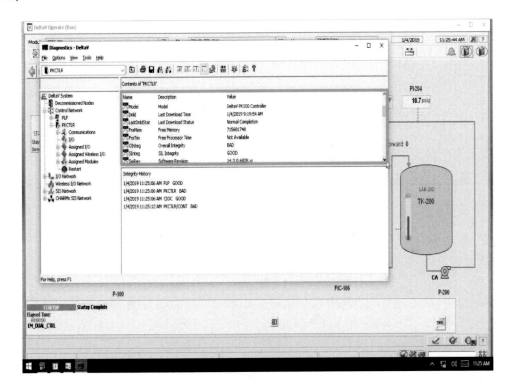

19.11   Section 6.6 Answers – Effect of Scan Rate on Control

Step 6.6.8

It took longer for the loop to reach the setpoint with the slower scan rate.

Graphic 6-6-8 DeltaV Scan Rate Response

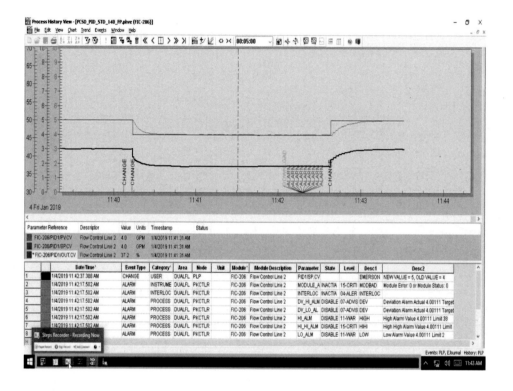

19.12   Section 6.7 Answers – Effect of Filters on Control

Step 6.7.6

Adding a filter slowed the control response and produced some overshoot.

Graphic 6-7-6 DeltaV Filter Response

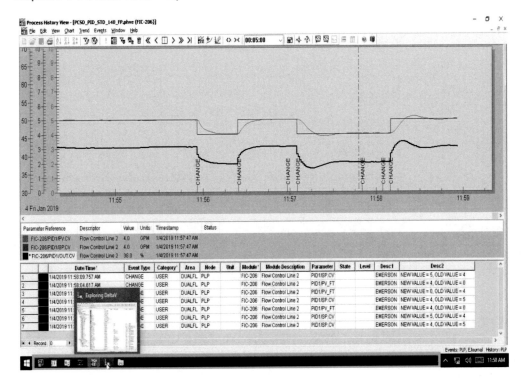

19.13   Section 7.5 Answers – Level Transmitter Accuracy

Step 7.5.1

Rosemount 5300 guided wave radar accuracy in water is +/- 0.12 inches.

Step 7.5.2

Rosemount 3051 differential pressure transmitter accuracy is +/- 0.075% of span.

Step 7.5.4

Rosemount 5300 guided wave radar transmitter started to respond at about 2 ounces of water.

Step 7.5.5

Rosemount 3051 differential pressure transmitter started to respond at about 1 ounce of water.

Step 7.5.6

With no flow, Rosemount 5300 guided wave radar was moving at about 0.01 percent.

Step 7.5.9

With flow, Rosemount 5300 guided wave radar was moving at about 0.03 percent.

19.14   Section 7.6 Answers – Level Transmitter Configuration via AMS

Step 7.6.1

LIT-103 is configured for 0 to 24 INWC.

Graphic 7-6-1 AMS Configuration Result

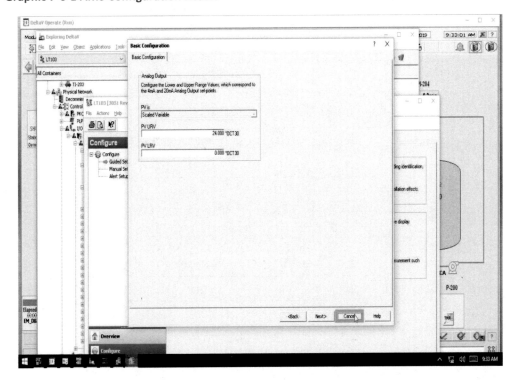

Step 7.6.5

A transmitter in ethanol service would be set up as URV = 24.0 INWC * 0.789 = 18.94 INWC. 0.0 to 18.94 INWC for ethanol service.

19.15 Section 7.7 Answers – Tank Strapping

Step 7.7.12

Data Record

Table 7-7-12 Tank Strapping

Actual Gallons Per Tank Embossed Scale	LIT-103 INWC	LI-103 DeltaV Percent	LIT-101 Level Feet	LIT-101 Distance Feet	LIC-101 DeltaV Percent
1.75	3.146	13.50	0.34	1.76	16.50
5	5.93	24.70	0.51	1.59	25.60
8	7.93	33.00	0.66	1.44	33.00
10	9.12	38.20	0.76	1.34	38.00
12	10.44	43.50	0.87	1.23	43.30
15	12.16	50.70	1.05	1.05	52.40
17	13.34	55.60	1.18	0.92	59.00
20	15.37	64.00	1.39	0.71	69.20
21	15.94	66.40	1.43	0.67	70.40
22	16.44	69.40	1.48	0.62	73.40
23	17.31	72.20	1.51	0.59	74.50
24	17.91	74.70	1.55	0.55	76.70
25	18.6	77.70	1.59	0.51	79.50

Step 7.7.15

We would expect the graph to be non-linear due to tank geometry.

Figure 7-7-15A LI-103 Gallons vs. Percent Level

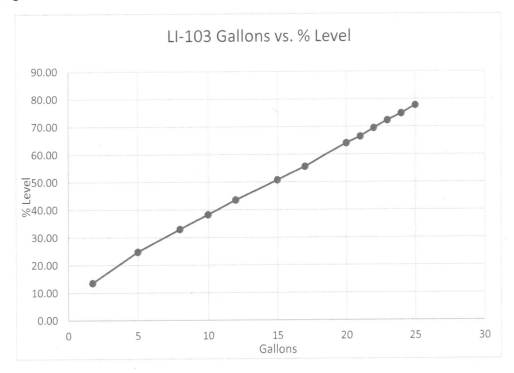

Figure 7-7-15B LIC-101 Gallons vs. Percent Level

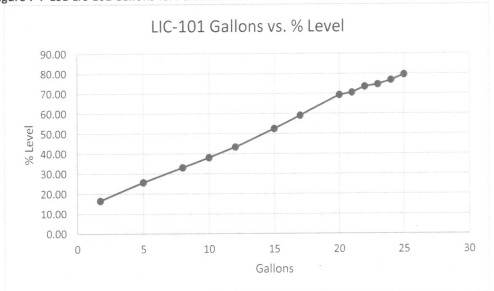

Step 7.7.16

The tank strapping produced a linear chart because the embossed tank markings did not really start to widen below 25 gallons. Given the fact thatTK-100 cross sectional area decreases as the elevation increases, we would expect the tank markings to increase sooner. When strapping a tank in a production environment, a NIST (National Institute of Standards and Technology) calibrated test standard should be used to record gallons.

Step 7.7.17

Process engineers are generally only concerned with the percent level when it comes to safety interlocks. For processing they generally want to know the volume in the tank. Based on the density, temperature and pressure, they can calculate the mass for a given volume. Theoretical chemical reaction calculations are done in mass which is why mass is so important. From an accounting standpoint, the accountants want to know how much raw material is in the storage tank so they can do work in progress calculations. This is an indication of how well production is going and identifies the need to order additional raw materials.

19.16 Section 8.5 Answers – Effect of Entrained Gases on Magnetic Flowmeter Operation

Step 8.5.6

Starting level in TK-100 was 23.1%.

Step 8.5.9

FIC-206 control was normal. It controlled within 0.1 GPM of setpoint.

Step 8.5.11

Final level in TK-100 was 55.8%.

Differential level was 32.7%.

Step 8.5.12

Starting level in TK-100 was 21.5%.

Final level in TK-100 was 46.1 %.

Differential level was 24.6%.

The control of FIC-206 appeared to be normal, within 0.2 GPM of setpoint, but the amount charged in 2 minutes was significantly different. The magnetic flowmeter must remain full during operation in order to give an accurate measurement. Entrained gases will give erroneous readings.

Step 8.5.16

Normal 5 Hz Signal to Noise Ratio = 202.9

Normal 37 Hz Signal to Noise Ratio = 1336.4

Step 8.5.17

5 Hz Signal to Noise Ratio with Entrained Gas = 195.1

37 Hz Signal to Noise Ratio with Entrained Gas = 274.3

The entrained air is seen as noise by the meter, so the signal to noise ratio values decreased. The higher the signal to noise strength is, the more accurate the meter will be. By switching to 37 Hz, the meter will do a better job dealing with the entrained air.

19.17 Section 8.6 Answers – Orifice Plate Differential Pressure Flow

8.6.2

Square Root is Enabled.

Graphic 8-6-2-1 AMS Square Root Menu Item

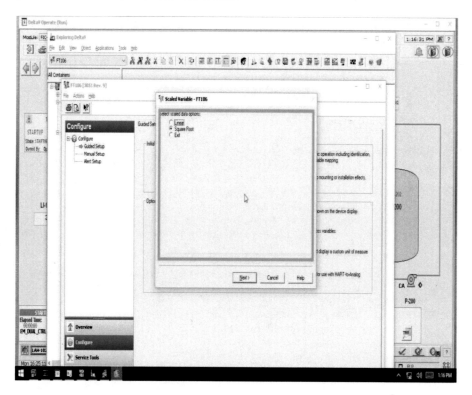

Upper Range in inH2O = 292.912994 inH2O

Graphic 8-6-2-2 AMS Upper Range inH2O

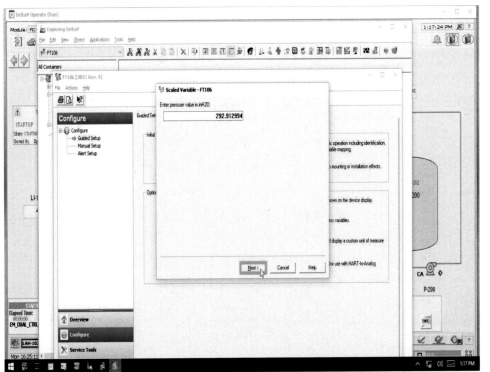

Upper Range GPM = 25 GPM

Graphic 8-6-2-3 AMS Upper Range GPM

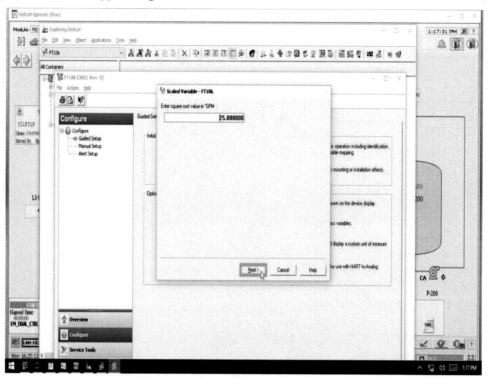

Step 8.6.5

Flow Rate = 5.746 GPM

Pressure = 15.49 inH2O

Graphic 8-6-5 AMS Readings Equalization Valve Closed

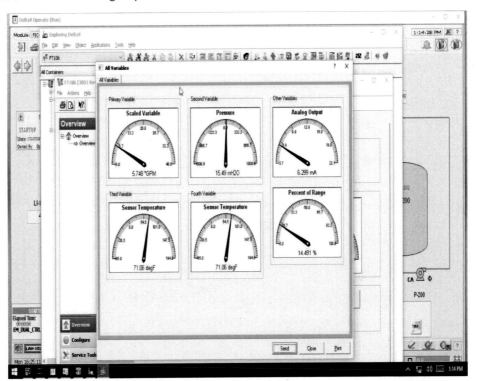

Step 8.6.6

Flow Rate = 4.391 GPM

Pressure = 9.04 inH2O

Once the equalization valve was opened, the differential pressure went down immediately, and the flow reading was incorrect. Eventually, the PLP would have shut down due to level since the cascade controller was acting on the wrong process input. Theoretically, the differential pressure would have gone to zero, but the reading we were seeing was due to sensor error since the sensor range was quite large.

Graphic 8-6-6 AMS Readings Equalization Valve Open

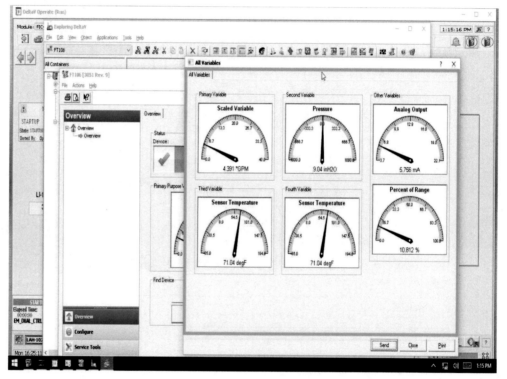

8.6.9 Orifice plate diameter = 0.680 inches

Figure 8-6-9 Orifice Plate Calculation

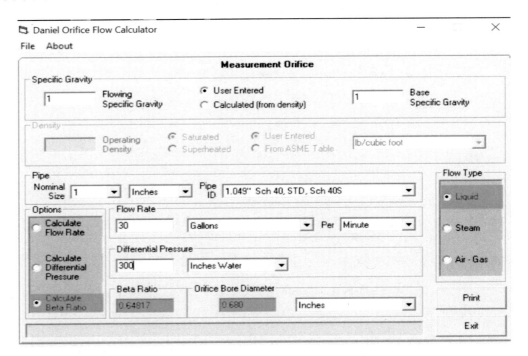

19.18   Section 8.7 Answers – Effect of Entrained Gas on Coriolis Flow Meter Operation

Step 8.7.3

Volumetric Flow Rate = 5.07 GPM

Mass Flow Rate = 42.21 lb/min

Density = 0.99797 g/Cucm

Step 8.7.4

Drive gain = 2.22%

Step 8.7.5

Volumetric Flow Rate = 2.81 GPM

Mass Flow Rate = 15.92 lb/min

Density = 0.49447 g/Cucm

Drive gain = 100.0%

When the drive gain gets to 100%, the meter loses its ability to accurately read the mass flow. There is too much noise due to the entrained air.

19.19 Section 8.8 Answers – Zeroing a Coriolis Meter

Step 8.8.3

Zero value before Zero Procedure = 0.003054 µs

Zero value after Zero Procedure = 1.497138 µs

No Flow Volumetric Flow Rate = -5.0 GPM

The zero calibration procedure made the 5 GPM flow rate the new zero.

Step 8.8.4

Factory Zero value = 0.000000µs

Step 8.8.5

Zero value = 0.004060µs

The new zero takes into account the process fluid and also any pipe stresses after the installation.

19.20 Section 8.9 Answers – Effect of Vortex Flowmeter Low Flow Cutoff Setting

Step 8.9.3

Low Flow Cutoff in Engineering Units = 3.361741 GPM

Low Flow Cutoff Frequency = 16.7 Hz

Recommended Low Flow Cutoff = 2.054190 GPM

Step 8.9.4

The low flow cutoff values remained the same.   The meter had previously be optimized.

Step 8.9.5

Lowest measurable flow rate = 3.4 GPM

Percent valve open at lowest measurable flow rate = 33% open

Step 8.9.6

Low Flow Cutoff in Engineering Units = 2.386213 GPM

Low Flow Cutoff Frequency= 10.0 Hz

Step 8.9.7

Lowest measurable flow rate = 2.4 GPM

Percent valve open at lowest measurable flow rate = 28% open

By manual manipulating the Low Flow Cutoff Frequency, we could get to meter to read lower flow rates than the optimal settings calculated.   There is a tradeoff; by doing this the meter is more susceptible to noise at lower flow rates.

19.21 Section 9.5 Answers – Zeroing a Pressure Transmitter

Step 9.5.3

Normal Operating Pressure PI-204 = 18.8 PSIG

Step 9.5.6

Atmospheric Pressure PI-204 = -0.029 PSIG

Step 9.5.10

Zeroed Pressure PI-204 = 0.01 PSIG

Step 9.5.11

Plug Replaced Pressure PI-204 = 0.435 PSIG

Pressure went up due to air being compressed when the plug was reinstalled.

Step 9.5.12

Drain Valve Closed Pressure PI-204 = 0.762 PSIG

Pressure went up due to air being compressed when the drain valve was closed.

Step 9.5.13

Isolation Valve Open Pressure PI-204 = 18.75 PSIG

Pressure returned to normal operating pressure in step 9.5.3.

Step 9.5.14

Since the pressure transmitter is installed above the line in liquid service, air is trapped between the transmitter and process fluid. Since air is compressible, this introduces a small amount of error into the measurement.

19.22 Section 10.5 Answers – Control Valve Failure

Step 10.5.2

The valve is fail closed and it slowly closes due to loss of air.

19.23 Section 10.6 Answers – Positioner Functionality

Step 10.6.3

In this example:

Setpoint % = 14.9%

Pressure A in PSIG = 19 PSI

Travel % = 14.9%

Graphic 10-6-3 Positioner Response With Obstruction

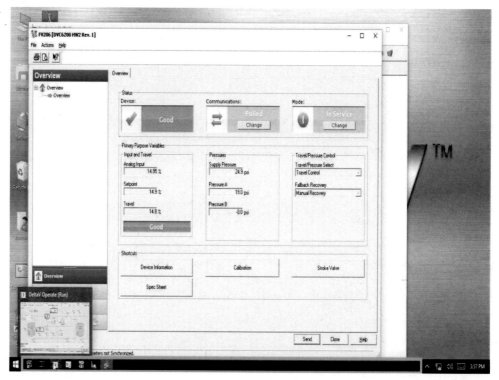

Step 10.6.5

In this example:

Setpoint % = 15%

Pressure A in PSIG = 6.5 PSI

This makes sense since valve is calibrated 5 to 15 PSI.   ((15-10) (0.15)) + 5 = 6.5 PSI

Travel % = 15.2%

Graphic 10-6-5 Positioner Response Without Obstruction

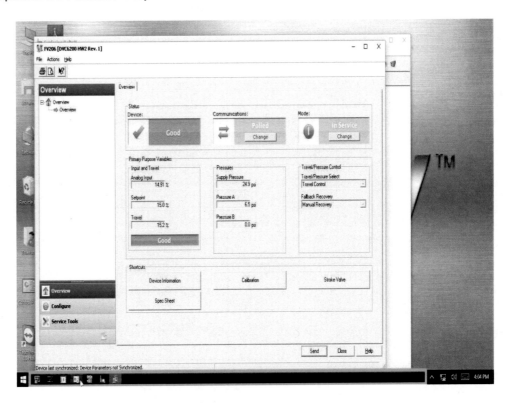

The amount of air pressure required to get to the target setpoint increased dramatically due to the obstruction interfering with the operation.

19.24   Section 10.7 Answers – Control Valve Response

Step 10.7.3

Table 10-7-3 Pressure and Flow Rates and Cv for FIC-206 at Different Valve Percentages

FV-206 % Open	PI-204 Pressure in PSIG P1	Pressure Drop across valve P1 -R assuming constant line losses	Flow of FIC-206 in GPM Q	Adjustment Factor F	Calculated $Cv = \frac{FQ}{\sqrt{P1-R}}$
0	-	-	0	-	0
5	20.1	5.6	0.4	0.2974	0.05
10	19.9	5.4	0.6	0.2974	0.08
15	19.5	5.0	1.1	0.2974	0.14
20	19.4	4.9	1.6	0.2974	0.21
25	19.1	4.6	2	0.2974	0.28
30	18.9	4.4	2.9	0.2974	0.41
35	18.8	4.3	3.6	0.2974	0.52
40	18.7	4.2	4.4	0.2974	0.64
45	18.6	4.1	5.5	0.2974	0.81
50	18.4	3.9	6.7	0.2974	1.01
55	18.1	3.6	8.4	0.2974	1.32
60	17.4	2.9	12.0	0.2974	2.10
65	16.7	2.2	13.4	0.2974	2.67
70	16.3	1.8	14.6	0.2974	3.24
75	15.7	1.2	15.9	0.2974	4.32
80	15.3	0.8	16.8	0.2974	5.58
85	15.1	0.6	17.1	0.2974	6.56
90	15.0	0.5	17.3	0.2974	7.27
95	14.9	0.4	17.4	0.2974	8.18
100	14.8	0.3	17.5	0.2974	9.5

R = 14.5 PSI

F = 0.297

Figure 10-7-3-1 FIC-206 Flow vs. FV-206 % Open

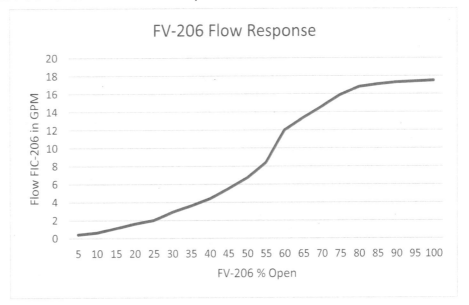

Figure 10-7-3-2 FV-206 % Open vs. Cv

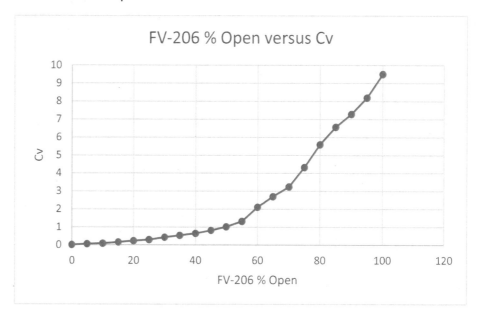

The installed flow response produced a somewhat linear response even though the valve is an equal percentage valve. This is due to the pump curve.

19.25 Section 11.5 Answers – I/O Identification

Step 11.5.2

Device Tag = FT106

Device Tag Parameter = FIELD_VAL_PCT

Other Device Tag Parameters = HART_DV_SLOT0, HART_DV_SLOT1, HART_DV_SLOT2, HART_DV_SLOT3, HART_FIELD_VALUE, HART_FV, HART_PV, HART_SV, HART_TV,

Graphic 11-5-1 Analog Input I/O Input Types

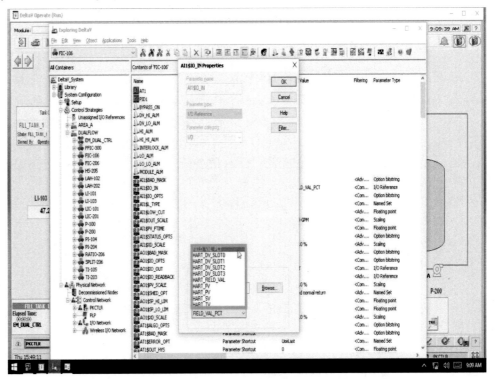

Step 11.5.3

FIC-106 Input Charm Address = CHM1-06

Step 11.5.4

FIC-106 Input Charm Type = 4-20 mA Analog Input with HART

19.26   Section 11.6 Answers – I/O Scaling

Step 11.6.2

FIC-206 Input Charm Address = CHM2-02

FIC-206 Input Charm Value = 12.53%

FIC-206 PV on DeltaV = 5.0 GPM

FIT-206 Transmitter reading = 5.0 GPM

Section 11.6.4

FIC-206 Input Charm Value = 12.46%

FIC-206 PV on DeltaV = 5.0 GPM

FIT-206 Transmitter reading = 6.2 GPM

They are different due to scaling. When the transmitter was re-spanned from 0 – 40 GPM to 0 - 50 GPM, the scaling configuration on FIC-206/AI1 was not changed along with it. The controller will still try and control at 5.0 GPM, but it is using the wrong input. The transmitter value is correct. Therefore, you need to verify the upper range limit on a transmitter during the commissioning phase. A simple milliamp simulate test is not enough to ensure the field transmitter and the control system are in sync.

19.27   Section 11.7 Answers – Analog Input Wiring

Step 11.7.1

2 Wire Transmitter = FIT-106.

4 Wire Transmitter = FIT-206.

Step 11.7.2

FIC-106 Input Charm Model Number = KL3021X1-BA1 AI 4-20mA HART.

FIC-206 Input Charm Model Number = KL3021X1-BA1 AI 4-20mA HART.

The Charm model numbers are the same.

Step 11.7.3

The terminations were the same on the prototype example PLP. Both transmitters were using terminals 1 and 2. The reason was that the magnetic flow meter is designed to support either loop powered or external powered wiring. Inside the wiring bay of FIC-206, the polarity was swapped. Normally, an externally powered 4 wire transmitter would use terminals 2 and 4.

Step 11.7.4

FIC-106 Charm Termination Block Model Number = KL4502X1-BA1

FIC-206 Charm Termination Block Model Number = KL4502X1-BA1

They are the same. The analog input terminal block supports both loop powered and external powered transmitters.

19.28   Section 12.5 Answers –Network Diagnostics

Section 12.5.1

PKCTLR Primary IP Address = 10.4.0.30

PKCTLR Primary Ping Time = < 1ms

Step 12.5.2

PKCTLR Secondary IP Address = 10.8.0.30

PKCTLR Secondary Ping Time = < 1ms

All devices on the primary highway have a 10.4.0.X address. All devices on the secondary highway have a 10.8.0.X address.

19.29   Section 12.6 Answers – Network Redundancy

Step 12.6.2

A yellow PLP alarm appears but the PLP keeps running.

Step 12.6.3

Everything on the main graphic is lost and turns magenta. The PLP keeps running.

Step 12.6.5

A yellow PKCTLR alarm appears but the PLP keeps running.

Step 12.6.6

A red COM PKCTLR alarm appears but the PLP keeps running. The CIOC and workstation are using their last values. This is where a hardwired ESTOP is needed because you have no way to shut down the process.

Step 12.6.7

When communications are resumed the controller responds to the loss of the real inputs and shuts the PLP down.

Step 12.6.8

A yellow CIOC alarm appears but the PLP keeps running.

Step 12.6.9

A red COM CIOC alarm appears but the PLP keeps running. The controller and CIOC are using their last values. The workstation is still updating with the controller data. This is where a hardwired ESTOP is needed because you have no way to shut down the process.

Step 12.6.10

When communications are resumed the controller responds to the loss of the real inputs and shuts the PLP down.

19.30   Section 13.5 Answers - Loose Wire

Step 13.5.2

The PLP shuts down and the Tank Operation sequence goes to Hold. FIC-206, FT-206 and EM_DUAL_CTRL alarms show up. The EM alarm alerts the operator that the equipment module went to hold. The process variable went to -10 GPM which is -125% of the PV span.

Step 13.5.3

The PV background color turned a burnt orange, indicating the transmitter is not communicating. The I/O Diagnostic detail indicates an IO Input Error.

Step 13.5.4

The hold monitor display indicates that the First Out Failure Go occurred due to module FIC-206.

Graphic 13-5-4 Equipment Module Failure Display

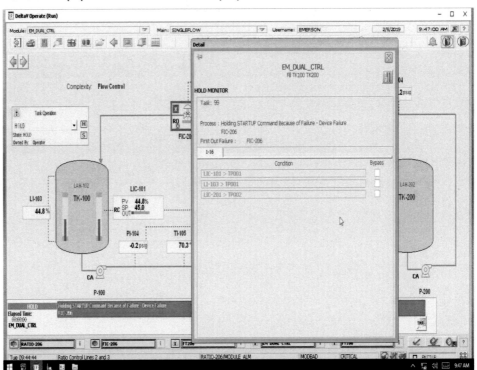

Step 13.5.6

Exact time the wire was disconnected: 2/5/2019 9:44:44.473 AM.

Graphic 13-5-6 Event Chronicle Alarm Capture

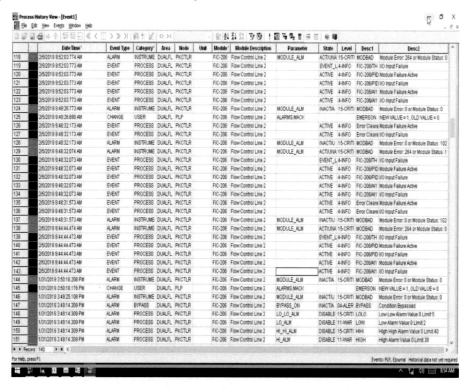

Step 13.5.7

Exact time the wire was disconnected: 2/5/19 9:44:44.473 AM. It matches the time captured in step 13.5.6.

Graphic 13-5-6 PHV Faceplate Event Capture

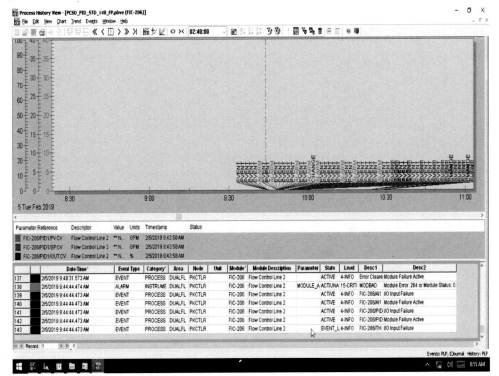

19.31   Section 13.6 Answers – Loss of Instrument Air

Step 13.6.2

The PLP eventually goes to Hold and shuts down due to low flow. The output of FIC-206 ramped full open, but the FV-206 actual position was closed.

Step 13.6.3

The First Out Failure was P-100.

Step 13.6.4

By clicking on the Interlock Detail for P-100, you can see the first out trip indicated by the arrow as "FV106 Flow < 3" which was caused by the loss of instrument air. Since the mode on P-100 is not LO, the pump is not interlocked. The low flow interlock is written with a delay and only is monitored if P-100 is running.

19.32 Section 13.7 Answers – Loss of Power

Step 13.7.2

The PLP went to Hold since P-200 failed. Alarms that appeared are LAH-102 and equipment module EM_DUAL_CTRL.

Step 13.7.3

P-200 was the First Out Failure.

Step 13.7.4

On the Interlock Detail of P-200, the first out arrow was pointing at LAH-102. P-200 is in LO mode and still interlocked. LAH-102 is configured and wired failsafe, so when the switch lost power, P-200 failed which put the equipment module in hold. Sometimes the root cause of the failure is not apparent.

Step 13.7.6

The PLP kept running. No alarms appeared. None of the I/O wired to Charms Baseplate 2 require external power to operate.

Step 13.7.8

A CIOC primary power alarm appear, but the PLP keeps running.

Step 13.7.9

The PLP goes to Hold and shuts down. The whole screen turned red since many modules lost communication with the field.

Step 13.7.10

The equipment module First Out Failure was P-100.

Step 13.7.11

FV-206 closed in the field but the controller output on DeltaV was held at its last value. FIT-206 read 0.0 GPM on the transmitter, but on DeltaV it read 5.0 GPM. When the CIOC stopped communicating with the controller, the controller held the last good value.

Step 13.7.12

A red X appears on the AI1/OUT parameter in Control Studio On-Line.

19.33 Section 14.5 Answers – Signal Cable Shielding

Step 14.5.5

A hair dryer was used as the AC load. In the trend below, the shield was re-terminated at 11:03:30. There may have been a little reduction in noise but it was insignificant. If a longer time span and a larger load were used, there would have been some noise due to a charge build up on the shield.

Graphic 14-5-5 PI-204 Noise Trend

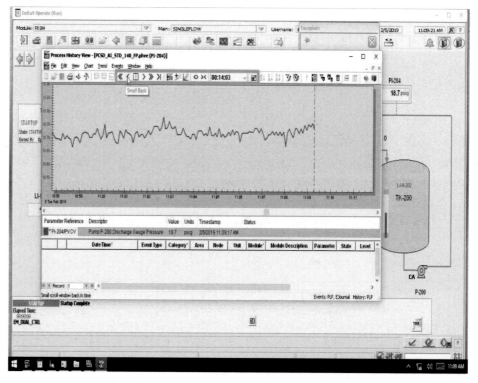

19.34   Section 14.6 Answers – AC and DC Grounding

Step 14.6.9

No change was observed on PI-204 with the power grounds disconnected. The system ground was floating but still referenced to the same point. With the AC power ground removed, personnel are unprotected if a ground fault occurred. Also, the system would likely see a spike on the input and output low voltage signals if a ground fault occurred.

Step 14.6.10

No change was observed on PI-204. In a normal system the AC and DC grounds would not be connected at this point to prevent power faults from causing noise on the DC signal cabling.

19.35 Section 15.5 Answers – Discrete Control Device Interlock

Step 15.5.2

The PLP shut down due to the LAH-102 high level as noted in the Hold monitor. The actual mode of HS-205 went to LO and was locked in that mode. The module would not accept setpoint changes. The first out trap on HS-205 was high level on LAH-102. The valve did not fail because the valve has no feedback (limit switches). It is an output only DCD so the PV tracks the setpoint at all times.

Step 15.5.3

The PLP would not start and due to the Hold condition of LAH-102 high level being in alarm. P-200 was interlocked due to LAH-102 in alarm.

Step 15.5.4

The valve remains in the passive State even after the interlock is cleared. A sequencer or an operator would need to open it back up again. Since the interlock has cleared, the mode on HS-205 can be changed to AUTO and the setpoint driven to OPEN.

19.36 Section 16.6 Answers – Controller Action

Step 16.6.2

The FIC-206 loop error is negative (PV – SP) and the output would need to increase to bring the loop back to setpoint. This is reverse action.

Step 16.6.3

The Loop Tuning Detail shows FIC-206 is setup as reverse.

Step 16.6.4

The LIC-101 error is negative and continues to increase due to the integrating loop. The output would need to decrease to bring the loop back to setpoint. This is direct action. The LIC-101 Loop Tuning Detail shows LIC-101 is setup as direct.

19.37   Section 16.7 Answers – Open Loop Controller Tuning

Step 16.7.7

$$R = \frac{(7.4-4.4)\left(\frac{100}{40}\right)}{3.3} = 2.27 \text{ %/sec}$$

L = 1.1 seconds

U (step change in percent) = 10%

Step 16.7.8

$$K_c = \frac{(0.9)\,(10)}{(1.1)(2.27)} = 3.6$$

$T_i = (3.3)\,(1.1) = 3.63$ sec

Step 16.7 9

$K_c = 0.75$

$T_i = 2$ sec

$T_d = 0$ sec

Step 16.7.10

Figure 16-7-10 Closed Loop Response with Default Tuning

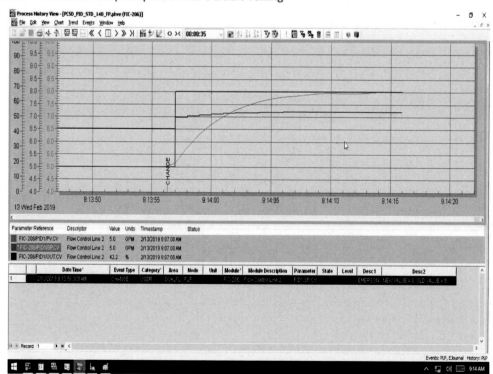

Step 16.7.12

The Ziegler-Nichols settings got to setpoint quicker on the first oscillation but had much more peak and integrated error. The default settings stabilized quicker and had less integrated error. The Ziegler-Nichols settings overshot setpoint as expected with quarter amplitude tuning.

Figure 16-7-12 FIC-206 Closed Loop Response with Ziegler-Nichols Open Loop Tuning Settings

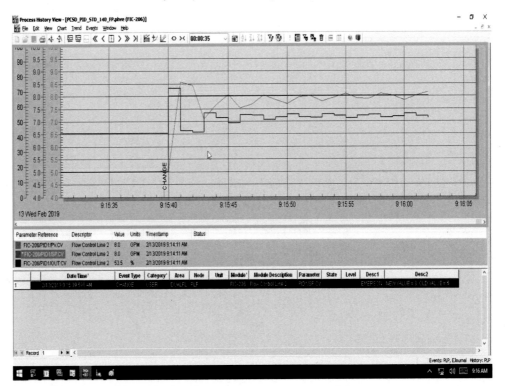

Step 16.7.14

θ_0 = 1.1 sec (open loop dead time from trend)

((7.0 - 4.4) (0.63)) + 4.4 = 6.038 GPM corresponding open loop time constant from trend T_i = 2.0 sec

$K_0 = \dfrac{(((7.0 - 4.4))(40))/(100))}{10.0} = 0.65$

$\lambda = (3)(1.1) = 3.3$ sec

$K_c = \dfrac{2}{(0.65)(3.3+1.1)} = 0.69$

Step 16.7.16

The default tuning settings got to setpoint quicker but neither had any overshoot. The default tuning was faster, taking only 13 versus 17 seconds. The Lambda tuning produced much less peak and integrated error than the Ziegler-Nichols.

Figure 16-7-16 FIC-206 Closed Loop Response With Lambda Tuning Settings

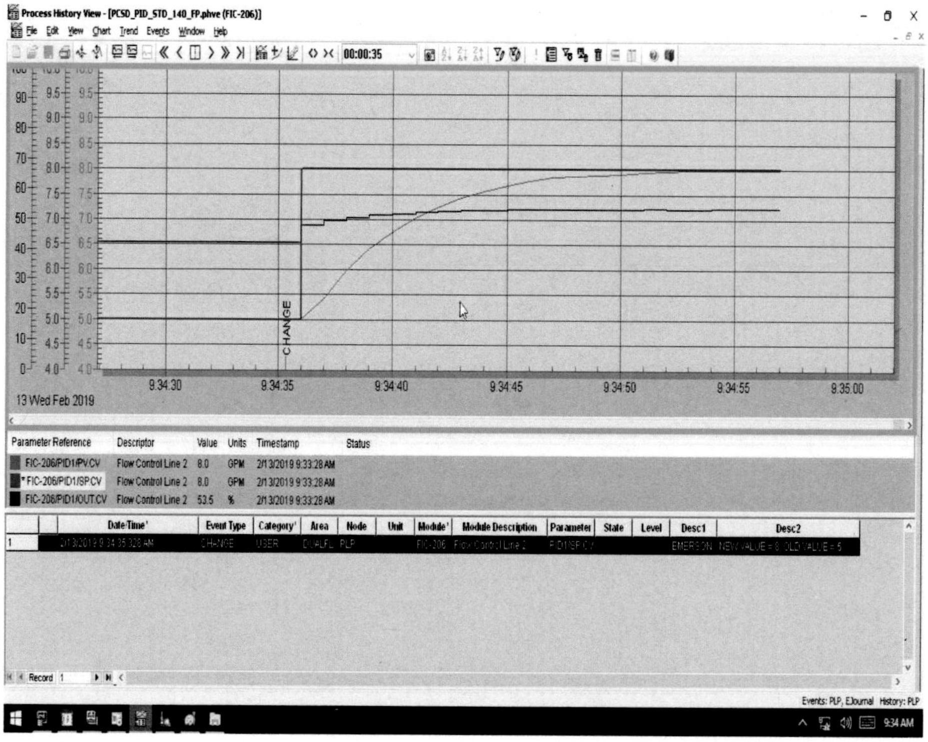

19.38   Section 17.6 Answers – Cascade Control Response

Step 17.6.1

K_c = 2.0

T_i = 120 sec

T_d = 0 sec

Step 17.6.7

Tuning was oscillatory. It was setup like a surge tank to minimize changes to downstream flow.

Figure 17-6-7 LIC-101 Closed Loop Cascade Response Default Tuning

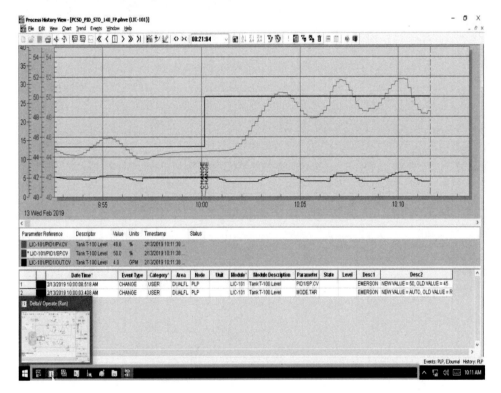

Step 17.6.10

Graphic 17-6-10 Cascade Open Loop Response

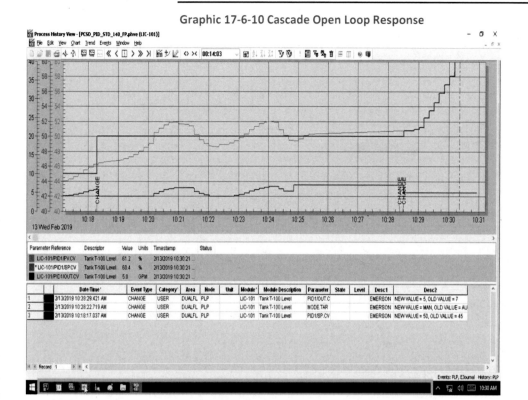

Step 17.6.11

θ_0 = 39 sec

λ = (3)(39) = 117 sec

$K_i = \dfrac{(54.5 - 51.7)/24}{5} = 0.023$ %/sec/%

$K_c = \dfrac{273}{(0.023)(117 + 39)^2} = 0.48$

T_i = (2)(117) + 39 = 273 sec

(0.5) (39) < T_d < (0.25) (273) = T_d selected = 50 sec

Step 17.6.13

The Lambda tuning closed loop response had less oscillation than the default tuning. The peak error and first oscillation time to setpoint was quicker.

Figure 17-6-13 LIC-101 Closed Loop Cascade Response With Lambda Tuning

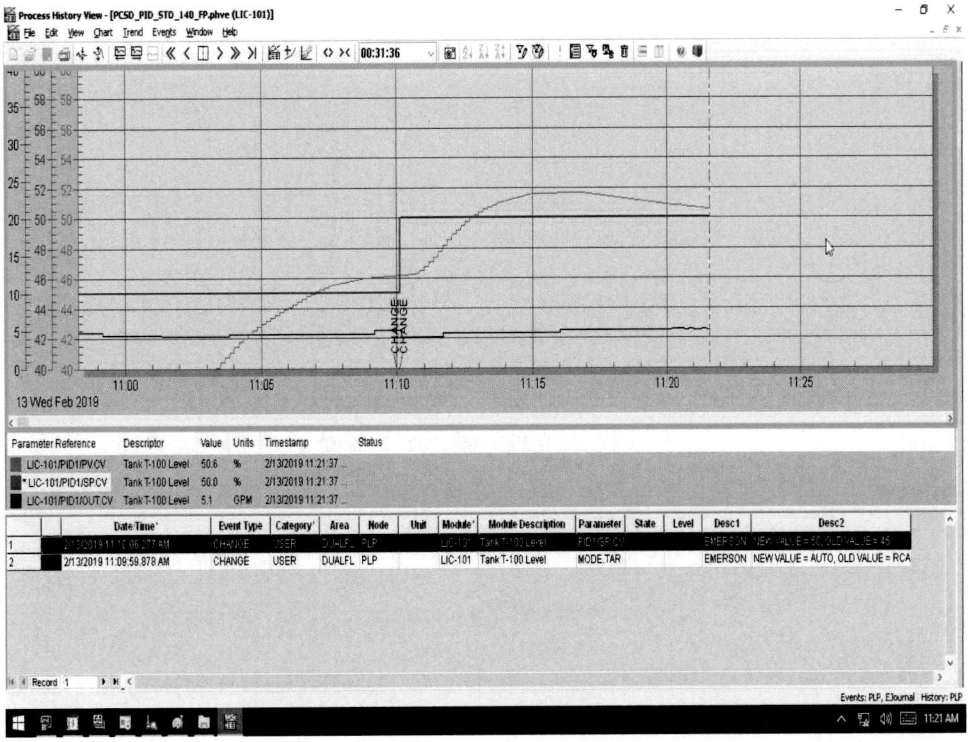

Step 17.6.21

Figure 17-6-21 LIC-101 Open Loop Response Cascade Bypassed

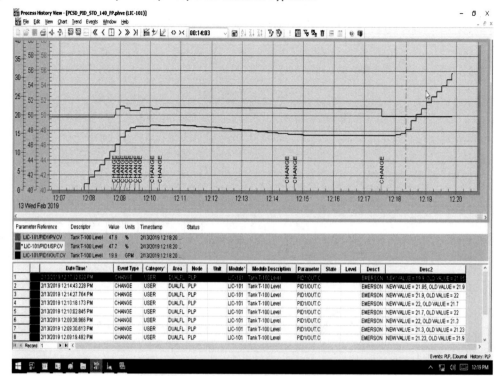

Step 17.6.22

θ_0 = 25 sec

λ = (3)(39) = 75 sec

$K_i = \dfrac{(56.0 - 51.0)/60}{5}$ = 0.0166 %/sec/%

$K_c = \dfrac{175}{(0.0166)(75+ 25)^2}$ = 1.05

T_i = (2)(75) + 25 = 175 sec

(0.5) (25) < T_d < (0.25) (175) = T_d selected = 30 sec

Step 17.6.24

The response with the cascade scheme took longer for LIC-101 to get to setpoint on the first oscillation but it had less overshoot. It seems the tuning on the cascade scheme can be more aggressive because the lower flow loop can filter out some disturbances.

Figure 17-6-22 LIC-101 Lambda Tuning Closed Loop Response Cascade Bypassed

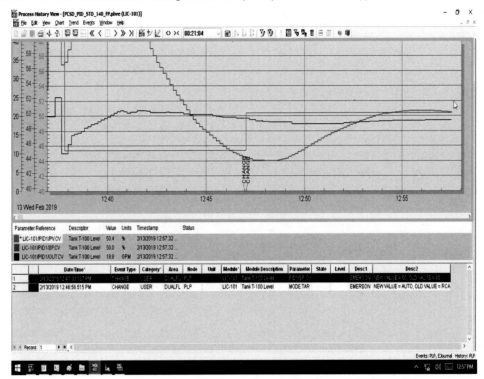

19.39   Section 17.7 Answers – Cascade Loop Signal Status Propagation

Step 17.7.3

FIC-106 PID1/BKCAL_OUT Status = GoodCascade

Step 17.7.4

LIC-101 PID1/BKCAL_IN Status = GoodCascade

Step 17.7.7

FIC-106 PID1/BKCAL_OUT status = Good Cascade High Limited

The loop status is high limited due to the valve being 100% open.

Step 17.7 8

LIC-101 PID1/BKCAL_IN status = Good Cascade High Limited

FIC-106 high limited status was propagated back to LIC-101 which reduced the contribution of the reset component to the LIC-101 PID response.

19.40   Section 18.5 Answers – Split-Range Control

Step 18.5.5

Table 18-5-5 SPLIT-206 Split Range Response

SPLIT-206 % Output	FV-206 % Open	FV-300 % Open
0	100	0
5	95	5
10	90	10
15	85	15
20	80	20
25	75	25
30	70	30
35	65	35
40	60	40
45	55	45
50	50	50
55	45	55
60	40	60
65	35	65
70	30	70
75	25	75
80	20	80
85	15	85
90	10	90
95	5	95
100	0	100

Step 18.5.6

The split-range controller was proportional. An example would be color control where one valve controlled the base, and the other valve controlled the pigment.

Appendix A: P&ID With Missing Components

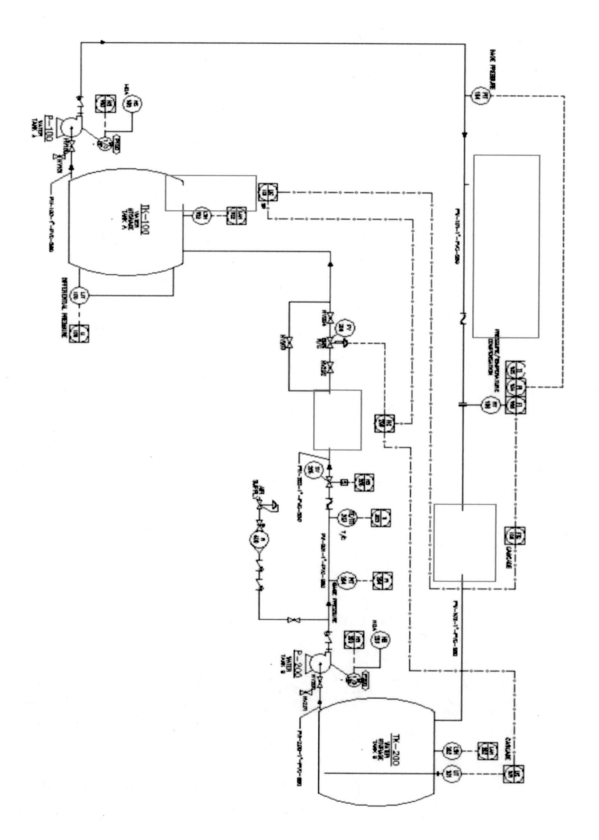

The contents of this publication are presented for information purposes only, and while effort has been made to ensure their accuracy, they are not to be construed as warranties or guarantees, express or implied, regarding the products or services described herein or their use or applicability. All sales are governed by our terms and conditions, which are available on request. We reserve the right to modify or improve the designs or specifications of our products at any time without notice.

Copyright Information

The Emerson logo is a trademark and service mark of Emerson Electric Co. PlantWeb, DeltaV, Rosemount, Fisher, FIELDVUE, AMS and Ovation are marks of one of the Emerson Process Management Family of companies. All other marks are the property of their respective owners.

Emerson@workforcedevelopment.com
www.emerson.com/en-us/automation/services-consulting/educational-services/performance-learning-platform

Please visit our website for up to date product data.
www.Emerson.com